GENOMIC APPROACHES IN EARTH AND ENVIRONMENTAL SCIENCES

New Analytical Methods in Earth and Environmental Science Series

Introducing New Analytical Methods in Earth and Environmental Science, a new series providing accessible introductions to important new techniques, lab and field protocols, suggestions for data handling and interpretation, and useful case studies.

New Analytical Methods in Earth and Environmental Science represents an invaluable and trusted source of information for researchers, advanced students, and applied earth scientists wishing to familiarize themselves with emerging techniques in their field.

All titles in this series are available in a variety of full-color, searchable e-book formats.

See below for the full list of books from the series.

GENOMIC APPROACHES IN EARTH AND ENVIRONMENTAL SCIENCES

GREGORY DICK

University of Michigan
Michigan, USA

WILEY Blackwell

Registered Office(s)
John Wiley & Sons, Inc., 111 River Street, Hoboken, NJ 07030, USA
John Wiley & Sons Ltd, The Atrium, Southern Gate, Chichester, West Sussex, PO19 8SQ, UK

Editorial Office
9600 Garsington Road, Oxford, OX4 2DQ, UK

For details of our global editorial offices, customer services, and more information about Wiley products visit us at www.wiley.com.

Wiley also publishes its books in a variety of electronic formats and by print-on-demand. Some content that appears in standard print versions of this book may not be available in other formats.

Library of Congress Cataloging-in-Publication Data has been applied for
9781118708248

Cover Design: Wiley
Cover Image: © cosmin4000/Gettyimages

Set in 10/12.5pt Minion by SPi Global, Chennai, India

Printed in Great Britain by TJ International Ltd, Padstow, Cornwall.

10 9 8 7 6 5 4 3 2 1

Contents

Preface

In recent years we have witnessed an explosion of DNA sequencing technologies that provide unprecedented insights into biology. Although this technological revolution has been driven by the biomedical sciences, it also offers extraordinary opportunities in the Earth and environmental sciences. In particular, the application of "omics" methods (genomics, transcriptomics, proteomics) directly to environmental samples offers exciting new vistas of complex microbial communities and their roles in environmental and geochemical processes. However, there is currently a lack of resources and infrastructure to educate and train geoscientists about the opportunities, approaches, and analytical methods available in the application of omic technologies to problems in the geosciences. This book aims to begin to fill this gap. Due to the rapidly advancing nature of DNA sequencing technologies, this book will almost certainly be well out of date by the time of publication. Nevertheless, my hope is that the accompanying e-book format will allow relatively frequent updates and will serve as a foundation and a gateway for students and other scientists to access this exciting field. I apologize in advance to the many researchers whose excellent work was inevitably not cited, due to either my own ignorance or constraints on space. I welcome suggestions for citations, additions, and corrections that can be incorporated into future editions.

Gregory Dick
Ann Arbor, Michigan, USA
August, 2017

Acknowledgments

This book is the product of many interactions with numerous people over several years. It was developed from lecture notes for a graduate class that I teach at the University of Michigan, Earth 523, and thus benefited from numerous questions, comments, and input from students during class over the years. Several students, including Shilva Shrestha and Matthew Hoostal, provided direct detailed comments and edits for which I am most grateful. Sunit Jain was a bioinformatician in my lab who assisted with and substantially contributed to Earth 523, thus he indirectly contributed to the content of this book as well, particularly Chapter 4. Other current and former lab members including Brett Baker, Karthik Anantharaman, and Sharon Grim also provided valuable material and feedback. Vincent Denef provided insightful feedback and edits to Chapter 2, especially the section on the ecological and evolutionary aspects of microbial genomes. Mike Wilkins, Mary Ann Moran, Frank Stewart, Mak Saito, Jake Waldbauer, Ann Pearson, Murat Eren, and Titus Brown also provided thoughtful comments and suggested edits on individual chapters. Illustrations were drafted by Stephanie O'Neil, an undergraduate student at the time. Chapters 1 and 3 drew on early material from drafts of an article now published in *Elements* magazine (Dick and Lam 2015, *Elements* 11: 403–408) with permission from the Mineralogical Society of America. I am grateful for permission to reuse this material as well as figures from other papers as described herein.

I owe several people thanks for their patience. First, the publisher, Wiley, including Ian Francis, Delia Sandford, Ramya Raghaven, and Sonali Melwani, for their assistance and for tolerating the tardiness of this book. Finally, thanks to my wife, Jenna, and kids, Adeline and Ben, for their support and patience.

Abbreviations

BAC	bacterial artificial chromosome
DNA	deoxyribonucleic acid
dnGASP	*de novo* genome assembly project
GAGE	genome assembly gold standard evaluations
GMG	geomicrobiology and microbial geochemistry
IGV	Integrative Genomics Viewer
mRNA	messenger RNA
OLC	overlap-layout-consensus
ORF	open reading frames
OTU	operational taxonomic unit
PCR	polymerase chain reaction
RAM	random access memory
RNA	ribonucleic acid
rrn	ribosomal RNA

1

Introduction

1.1 Exploring the Microbial World

Microorganisms shaped the geochemical evolution of our planet throughout its history, and they continue to play a key role in the modern world. In deep time they oxygenated Earth's atmosphere and set the stage for life as we know it. Today, microbes mediate global biogeochemical cycles, influence the speciation and fate of pollutants, and modulate climate change through production and consumption of greenhouse gases. The field of geomicrobiology and microbial geochemistry (GMG), which studies the interplay between microbes and the Earth system, has roots in the 19th century (Druschel & Kappler 2015; Druschel et al. 2014). However, only recently has the breadth of microbial geomicrobiological processes and extent to which they shape geological, geochemical, and environmental processes become clear. Many methods and concepts central to GMG are also relevant to environmental engineering (e.g., drinking water and wastewater treatment) and medicine (e.g., human microbiome), including the omics approaches that are the focus of this book.

How to study this microbial world? Inherent challenges abound; microorganisms are small. Their cellular morphology is typically not informative of their phylogeny, physiology, or role in biogeochemical or ecological processes. Microbes often live in highly diverse microbial communities where it is hard to decipher the activities of different microorganisms or to trace specific microbial processes. Traditional microbiological approaches revolve around the cultivation of bacteria and archaea, which enables powerful laboratory-based methods of dissecting microbial physiology, biochemistry, and genetics as they relate to geochemical processes (Newman et al. 2012). Yet most microorganisms in nature are resistant to cultivation owing to symbiotic lifestyles or unknown nutritional requirements (Staley & Konopka 1985). Further, it can

Genomic Approaches in Earth and Environmental Sciences, First Edition. Gregory Dick.
© 2019 John Wiley & Sons Ltd. Published 2019 by John Wiley & Sons Ltd.

be impractical to grow pure cultures due to the extremely slow growth of many microorganisms, which in the environment is perhaps more akin to stationary phase than to growing cultures (Roy et al. 2012). Comprehensive culturing is also impractical because of the stunning complexity of natural microbial communities (thousands of species). Finally, the results from pure cultures may not be representative of *in situ* processes (Madsen 2005).

Traditional geochemical methods of measuring process rates and products and using biological poisons or inhibitors of specific microbial enzymes offer critical quantitative data and some mechanistic insights (Madsen 2005; Oremland et al. 2005). However, these approaches provide little information with regard to the identity or nature of the microorganisms that underpin processes of interest. Exciting advances in microscopy and spectroscopy that provide opportunities to link microorganisms to biogeochemical processes are described and reviewed elsewhere (Behrens et al. 2012; Newman et al. 2012; Wagner 2009).

Recent advances in DNA sequencing technologies open up entirely new avenues to study geomicrobiology by circumventing the cultivation step and providing extensive information on microorganisms as they exist in natural settings. This data comes from the sequence of macromolecules (Box 1.1)

Box 1.1 Definitions of key macromolecules studied by omics approaches

Deoxyribonucleic acid (DNA): DNA consists of four nucleotide bases – guanine (G), adenine (A), thymine (T), and cytosine (C) – that are joined together in a sequence to form genes.

Gene: a unit of genetic information encoding protein, tRNA, or ribosomal RNA. Genes are about 1000 bases long, on average.

Genome: the genome is the collection of all genetic information in an organism, including the genes as well as elements between genes that are involved in regulating gene expression. Microbial genomes range in size from approximately 400 000 to 10 million bases and from 400 to 10 000 genes.

Ribonucleic acid (RNA): There are several major forms of RNA, including messenger RNA (mRNA), transfer RNA (tRNA), and ribosomal rRNA (rRNA). mRNA is an intermediate between DNA and protein (see Fig. 1.1); rRNA is a structural and catalytic component of ribosomes, the machinery that translates mRNA into protein. tRNA are small molecules that recognize the three-base code of mRNA and translate it into amino acids during protein synthesis.

Protein: proteins are polymers (long chains) of amino acids. The two main roles of proteins are (1) to provide structure or scaffolding, e.g., in cell wall or protein synthesis; (2) to catalyze biochemical reactions in the cell, including those required for energy metabolism, biosynthesis of macromolecules, transport of elements into and out of the cell, and generation of biogenic minerals ("biomineralization"). Proteins can also "sense" the environment and transduce signals that elicit cellular responses.

Lipids: hydrocarbons, often with polar head groups, that are the primary constituents of cell membranes. In some cases, specific lipids are diagnostic of specific microbial groups or metabolisms. Unlike other biological macromolecules, lipids may be preserved in sediments over geological time (millions to billions of years), so they have great value in potentially providing information on ancient ecosystems. Like other macromolecules, the synthesis of lipids is conducted by proteins that are encoded by genes. Hence, the "lipidome" can theoretically be predicted from the genome.

Carbohydrates: macromoleucles consisting of carbon, hydrogen. Carbohydrates decorate the cell surface and are an important interface between the cells and their environment. Because they are often negatively charged, they can play important roles in binding cations and influencing biomineralization.

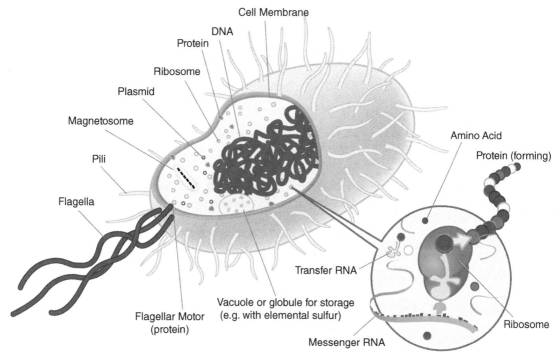

Figure 1.1 Generalized structure of a bacterial or archaeal cell. Inset details translation and protein synthesis. Source: Druschel and Kappler (2015), p. 390, Fig. 1. Reproduced with permission from the Mineralogical Society of America.

that constitute microbial cells (Fig. 1.1). This book focuses on DNA, RNA, and protein, and also touches on lipids and the pool of small molecules within a cell (metabolites). The collection of genes that encode an organism is known as the genome. Genes are transcribed as messenger RNAs, or transcripts, the total pool of which is called the transcriptome. Transcripts are then translated into protein, which actually performs the structural and biochemical functions of the cell. The total protein content of a cell is known as the proteome. The total content of small molecules within a cell is referred to as the metabolome. These small molecules include metabolites, the substrates, intermediates, and products of biochemical reactions catalyzed by enzymes. The study of the whole collection of each of these molecules in a pure culture is referred to as genomics, transcriptomics, proteomics, and metabolomics. When such information is derived from a whole community of microorganisms, we say "community genomics" or "metagenomics" (or metatranscriptomics, metaproteomics). Collectively, these approaches, whether applied to a single organism or a community of organisms, are referred to in shorthand as "omics."

Whereas genomes encode all the proteins that could possibly be made in a given cell, a genome does not give any information about which proteins and RNA are actually being produced at any given time, or about the quantities in which they are produced. Transcriptomics and proteomics provide

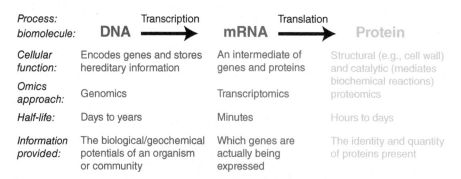

Figure 1.2 Macromolecules that serve as the basis for the three main omics approaches. Source: Dick and Lam (2015), p. 404, Fig. 1, with permission from the Mineralogical Society of America.

this information. DNA, RNA, and protein have different lifetimes based on the stability of the molecules and the biochemical mechanisms that degrade them. Thus these molecules provide information at different time scales (Fig. 1.2). Genomes also provide a "molecular fossil record" of how genes and organisms have evolved over the billions of years of life on Earth (David & Alm 2011; Macalady and Banfield 2003; Zerkle et al. 2005).

1.2 The DNA Sequencing Revolution: Historical Perspectives

The "meta-omics" revolution has its roots in the pioneering work of Carl Woese and colleagues, who sequenced microbial rRNA genes in order to uncover their phylogenetic relationships (Woese & Fox 1977). This work recognized that, because rRNA genes serve critical functions, they are present in every organism and are highly conserved at the sequence level. Thus, they hold invaluable information about the evolutionary relationships of microorganisms. Through painstaking labor, the sequence of rRNA genes from a wide range of organisms was deciphered, leading to an astonishing discovery: methane-producing microorganisms previously assumed to be bacteria were actually a new and completely separate domain of life – the archaea (Sapp & Fox 2013). This transformed our understanding of the tree of life by revealing that it is composed of three domains: bacteria, archaea, and eukarya (Woese & Fox 1977). The advent of rRNA gene sequencing also provided a practical and objective tool for classifying microorganisms, a task which had been declared impossible previously (Woese & Goldenfeld 2009).

Soon after, Pace and colleagues applied sequencing to rRNA genes purified directly from uncultured communities of microorganisms (Stahl et al. 1984). Subsequent application of polymerase chain reaction (PCR) to the amplification of rRNA genes (with an explicit focus on one of these genes, known as 16S rRNA) directly from the environment increased the throughput of this approach and revealed startling insights into the microbial world in seawater and other environments (DeLong 1992; Fuhrman et al. 1992). Spurred by rapidly advancing technologies and the ever declining costs and increasing throughput of DNA sequencing technologies (Loman et al. 2012), the culture-independent approach quickly revealed the staggering diversity of the microbial world (Pace 2009). This work revealed that only a tiny fraction of microbial groups have been studied in culture (Baker & Dick 2013; Pace 2009).

In parallel with the explosion of 16S rRNA gene sequencing, faster, cheaper DNA sequencing also enabled a new era of sequencing whole microbial genomes (Land et al. 2015). Information on the complete gene content theoretically provides a picture of the metabolic and physiological potential of microorganisms (however, see the caveats and challenges discussed in Chapter 3). The first bacterial genomes were published in 1995 (Fleischmann et al. 1995; Fraser et al. 1995), and the number of microbial genomes sequenced has expanded exponentially ever since (Fournier et al. 2013).

A major initial finding of these sequencing efforts was that microbial genomes have startling variability of gene content (Tettelin et al. 2005; Welch et al. 2002). This led to concepts of the pangenome, core genome, and flexible genome (Cordero & Polz 2014) (see Chapter 2). Genome sequences from cultured organisms are valuable because they enable studies of the links between genotype and phenotype and represent taxonomic and functional anchors in the tree of life for interpreting metagenomic data. Particularly valuable are genomes from type strains that have been validly described and named, which are estimated to account for a substantial portion (~15%) of phylogenetic diversity (Kyrpides et al. 2014). However, despite the microbial genome sequencing revolution, less than 3% of these type strains have had their genomes sequenced (Kyrpides et al. 2014). Thus, even the genomic coverage of *cultured* microbial life remains woefully inadequate, and of course, the cultured portion is just a small fraction of the total microbial world. The Microbial Earth Project (www.microbial-earth.org/) was recently launched to track the inventory of type strains of bacteria and archaea and their genome sequencing projects.

At the confluence of environmental 16S rRNA gene sequencing of microbial communities and whole genome sequencing of cultured microbes is the direct retrieval of genomes from uncultured microbial communities. Early metagenomic approaches used cloning of environmental DNA followed by sequencing and/or screening of expressed products for functions of interest (Riesenfeld et al. 2004; Stein et al. 1996). The term "metagenomics" was first coined in 1998, in the context of accessing natural products (e.g., antibiotics) from uncultured soil microorganisms (Handelsman et al. 1998). The power of the functional metagenomics approach lies in the direct connection of sequence to function and was illustrated beautifully by the discovery of bacterial light-driven proton pumps as a new form of phototrophy in the oceans (Béjà et al. 2000). This method can also provide valuable insights by directly linking phylogenetic marker genes to function (Pham et al. 2008), which is particularly valuable when the cloned fragments are large, as in BAC or fosmid libraries. However, because of the cost and labor involved in constructing and screening such clone libraries, this approach was not readily scalable. The "functional metagenomic" approach also faces practical challenges such as genetic and biochemical incompatibility between environmental genes and hosts (e.g., differences in codon bias, required cofactors). Some of these issues can be overcome by recent synthetic genomic approaches, but they still limit the throughput of exploratory, discovery-driven functional screening.

Shotgun metagenomics, in which community DNA is randomly fragmented and sequenced, was then demonstrated as a viable and valuable approach (Tyson et al. 2004; Venter et al. 2004) and quickly emerged as the dominant method used in metagenomics studies. For the first time, whole genomes of uncultured organisms could be reconstructed from microbial communities, revealing their metabolic potential (Tyson et al. 2004) and evolutionary processes (Allen & Banfield 2005). Several spectacular discoveries, including the linking of ammonia oxidation to archaea (Venter et al. 2004), demonstrated the power and promise of metagenomics. A vision for the potential advances that metagenomics could bring to science and society was beginning to come into view (National Research Council 2007). Hugenholtz and Tyson (2008) recount a brief history and highlights of these early stages and different approaches of metagenomics. For a more in-depth historical account see Handelsman (2004) and Gilbert and Dupont (2011). The rapid decrease in costs and increase in throughput of DNA sequencing has enabled shotgun sequencing of more complex microbial communities (Fig. 1.3). Recent papers report the reconstruction of thousands of genomes from metagenomes (Anantharaman et al. 2016).

While the genomic sequence provides information on the metabolic and physiological *potential* of microorganisms, it does not indicate whether those functions are being carried out at a particular point in time or space. To address this question, characterizing the expression of mRNA or protein is required. Metatranscriptomics was applied with great success to surface seawater microbial communities, revealing that flexible genes are highly expressed in the environment (Frias-Lopez et al. 2008). Critically, this paper also used qPCR to independently evaluate the accuracy of RNA amplification, which is required to obtain sufficient cDNA for many sequencing applications (see Chapter 9).

Whereas the DNA- and RNA-based analyses described above rely on the sequencing of nucleotides, proteomic tools use mass spectrometry to accurately measure the masses of small peptide fragments and even individual amino acids. The matching of these measured masses with calculated peptide masses derived from genomic information enables the identification of protein fragments. Metaproteomics is challenging because analytical methods for translating MS/MS spectra into protein sequence are complex

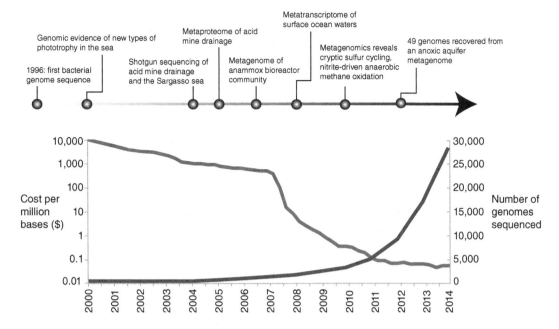

Figure 1.3 Major milestones in microbial community omics (*top*) and the decreasing cost and increasing throughput of DNA sequencing (*bottom*).

Source: Modified from Dick and Lam (2015), p. 406, Fig. 3, permission from the Mineralogical Society of America.

and largely reliant on having the corresponding genomic sequence for interpretation. Similarly, the recovery of total protein from many environmental samples is more challenging than the extraction of DNA and RNA. Not surprisingly, initial progress on application of proteomics to microbial communities was accomplished in low-diversity communities for which genomic sequence was available (Ram et al. 2005; Verberkmoes et al. 2009). Indeed, with sufficient genomic information, protein expression from very closely related strains can be differentiated (Lo et al. 2007). These studies yielded insights into the biochemical mechanisms of iron oxidation, a central process sustaining primary production and pyrite dissolution in acid mind drainage, and showed that among the most highly expressed proteins are "hypothetical" and "conserved hypothetical" proteins (Ram et al. 2005). With growing databases of genomic sequence and improving algorithms for interpreting MS/MS spectra, metaproteomics is now a viable approach for studying more complex microbial communities (see Chapter 10).

References

Allen, E. E. & Banfield, J. F. (2005) Community genomics in microbial ecology and evolution. *Nature Reviews Microbiology*, **3**, 489–498.

Anantharaman, K., Brown, C. T., Hug, L. A., et al. (2016) Thousands of microbial genomes shed light on interconnected biogeochemical processes in an aquifer system. *Nature Communications*, **7**, 13219.

Baker, B. J. & Dick, G. J. (2013) Omic approaches in microbial ecology: charting the unknown. *Microbe*, **8**, 353–360.

Behrens, S., Kappler, A. & Obst, M. (2012) Linking environmental processes to the in situ functioning of microorganisms by high-resolution secondary ion mass spectrometry (NanoSIMS) and scanning transmission X-ray microscopy (STXM). *Environmental Microbiology*, **14**, 2851–2869.

Béjà, O., Aravind, L., Koonin, E. V., et al. (2000) Bacterial rhodopsin: evidence for a new type of phototrophy in the sea. *Science*, **289**, 1902–1906.

Cordero, O. X. & Polz, M. F. (2014) Explaining microbial genomic diversity in light of evolutionary ecology. *Nature Reviews Microbiology*, **12**, 263–273.

David, L. A. & Alm, E. J. (2011) Rapid evolutionary innovation during an Archaean genetic expansion. *Nature*, **469**, 93–96.

Delong, E. F. (1992) Archaea in coastal marine environments. *Proceedings of the National Academy of Sciences of the United States of America* **89**, 5685–5689.

Dick, G. J. & Lam, P. (2015) Omics approaches to microbial geochemistry. *Elements*, **11**, 403–408.

Druschel, G. K. & Kappler, A. (2015) Geomicrobiology and microbial geochemistry. *Elements*, **11**, 389–394.

Druschel, G. K., Dick, G. J. & Boyd, E. S. (2014) Geomicrobiology and Microbial Geochemistry 2014 Workshop Report. Available at: https://dx.doi.org/10.6084/m9.figshare.3083524.v1 (accessed 25 October 2017).

Fleischmann, R. D., Adams, M. D., White, O., et al. (1995) Whole-genome random sequencing and assembly of Haemophilus influenzae Rd. *Science*, **269**, 496–512.

Fournier, P. E., Drancourt, M., Colson, P., Rolain, J. M., La Scola, B. & Raoult, D. (2013) Modern clinical microbiology: new challenges and solutions. *Nature Reviews Microbiology*, **11**, 574–585.

Fraser, C. M., Gocayne, J. D., White, O., et al. (1995) The minimal gene complement of Mycoplasma–Genitalium. *Science*, **270**, 397–403.

Frias-Lopez, J., Shi, Y., Tyson, G. W., et al. (2008) Microbial community gene expression in ocean surface waters. *Proceedings of the National Academy of Sciences of the United States of America* **105**, 3805–3810.

Fuhrman, J. A., Mccallum, K. & Davis, A. A. (1992) Novel major archaebacterial group from marine plankton. *Nature*, **356**, 148–149.

Gilbert, J. A. & Dupont, C. L. (2011) Microbial metagenomics: beyond the genome. *Annual Review of Marine Science*, **3**, 347–371.

Handelsman, J. (2004) Metagenomics: application of genomics to uncultured microorganisms. *Microbiology and Molecular Biology Reviews*, **68**, 669–685.

Handelsman, J., Rondon, M. R., Brady, S. F., Clardy, J. & Goodman, R. M. (1998) Molecular biological access to the chemistry of unknown soil microbes: a new frontier for natural products. *Chemistry and Biology*, **5**, R245–259.

Hugenholtz, P. & Tyson, G. W. (2008) Metagenomics. *Nature*, **455**, 481–483.

Kyrpides, N. C., Hugenholtz, P., Eisen, J. A., et al. (2014) Genomic encyclopedia of bacteria and archaea: sequencing a myriad of type strains. *Plos Biology*, **12**, e1001920.

Land, M., Hauser, L., Jun, S. R., et al. (2015) Insights from 20 years of bacterial genome sequencing. *Functional and Integrative Genomics*, **15**, 141–161.

Lo, I., Denef, V. J., Verberkmoes, N. C., et al. (2007) Strain-resolved community proteomics reveals recombining genomes of acidophilic bacteria. *Nature*, **446**, 537–541.

Loman, N. J., Constantinidou, C., Chan, J. Z. M., et al. (2012) High-throughput bacterial genome sequencing: an embarrassment of choice, a world of opportunity. *Nature Reviews Microbiology*, **10**, 599–606.

Macalady, J. & Banfield, J. F. (2003) Molecular geomicrobiology: genes and geochemical cycling. *Earth and Planetary Science Letters*, **209**, 1–17.

Madsen, E. L. (2005) Identifying microorganisms responsible for ecologically significant biogeochemical processes. *Nature Reviews Microbiology*, **3**, 439–446.

National Research Council (2007) *The New Science of Metagenomics: Revealing the Secrets of Our Microbial Planet*. National Academies Press, Washington DC.

Newman, D. K., Orphan, V. J. & Reysenbach, A. L. (2012) Molecular biology's contributions to geobiology. In: A.H. Knoll & K.O. Konhauser (eds), *Fundamentals of Geobiology*. Blackwell Publishing, Chichester.

Oremland, R. S., Capone, D. G., Stolz, J. F. & Fuhrman, J. (2005) Whither or wither geomicrobiology in the era of 'community metagenomics'? *Nature*, **3**, 572–578.

Pace, N. R. (2009) Mapping the tree of life: progress and prospects. *Microbiology and Molecular Biology Reviews*, **73**, 565–576.

Pham, V. D., Konstantinidis, K. T., Palden, T. & Delong, E. F. (2008) Phylogenetic analyses of ribosomal DNA-containing bacterioplankton genome fragments from a 4000 m vertical profile in the North Pacific Subtropical Gyre. *Environmental Microbiology*, **10**, 2313–2330.

Ram, R. J., Verberkmoes, C., Thelen, M. P., et al. (2005) Community proteomics of a natural microbial biofilm. *Science*, **308**, 1915–1920.

Riesenfeld, C. S., Goodman, R. M. & Handelsman, J. (2004) Uncultured soil bacteria are a reservoir of new antibiotic resistance genes. *Environmental Microbiology*, **6**, 981–989.

Roy, H., Kallmeyer, J., Adhikari, R. R., Pockalny, R., Jorgensen, B. B. & D'hondt, S. (2012) Aerobic microbial respiration in 86-million-year-old deep-sea red clay. *Science*, **336**, 922–925.

Sapp, J. & Fox, G. E. (2013) The singular quest for a universal tree of life. *Microbiology and Molecular Biology Reviews*, **77**, 541–550.

Stahl, D. A., Lane, D. J., Olsen, G. J. & Pace, N. R. (1984) Analysis of hydrothermal vent–associated symbionts by ribosomal-RNA sequences. *Science*, **224**, 409–411.

Staley, J. T. & Konopka, A. (1985) Measurement of in situ activities of nonphotosynthetic microorganisms in aquatic and terrestrial habitats. *Annual Reviews of Microbiology*, **39**, 321–346.

Stein, J. L., Marsh, T. L., Wu, K. Y., Shizuya, H. & Delong, E. F. (1996) Characterization of uncultivated prokaryotes: isolation and analysis of a 40-kilobase-pair genome fragment from a planktonic marine archaeon. *Journal of Bacteriology*, **178**, 591–599.

Tettelin, H., Masignani, V., Cieslewicz, M. J., et al. (2005) Genome analysis of multiple pathogenic isolates of Streptococcus agalactiae: implications for the microbial 'pan-genome'. *Proceedings of the National Academy of Sciences of the United States of America*, **102**, 16530–16530.

Tyson, G. W., Chapman, J., Hugenholtz, P., et al. (2004) Community structure and metabolism through reconstruction of microbial genomes from the environment. *Nature*, **428**, 37–43.

Venter, J. C., Remington, K., Heidelberg, J. F., et al. (2004) Environmental genome shotgun sequencing of the Sargasso Sea. *Science*, **304**, 66–74.

Verberkmoes, N. C., Denef, V. J., Hettich, R. L. & Banfield, J. F. (2009) Systems biology functional analysis of natural microbial consortia using community proteomics. *Nature Reviews Microbiology*, **7**, 196–205.

Wagner, M. (2009) Single-cell ecophysiology of microbes as revealed by Raman microspectroscopy or secondary ion mass spectrometry imaging. *Annual Review of Microbiology*, **63**, 411–429.

Welch, R. A., Burland, V., Plunkett, G. 3rd, et al. (2002) Extensive mosaic structure revealed by the complete genome sequence of uropathogenic Escherichia coli. *Proceedings of the National Academy of Sciences of the United States of America*, **99**, 17020–17024.

Woese, C. R. & Fox, G. E. (1977) Phylogenetic structure of the prokaryotic domain: the primary kingdoms. *Proceedings of the National Academy of Sciences of the United States of America*, **74**, 5088–5090.

Woese, C. R. & Goldenfeld, N. (2009) How the microbial world saved evolution from the Scylla of molecular biology and the Charybdis of the modern synthesis. *Microbiology and Molecular Biology Reviews*, **73**, 14–21.

Zerkle, A. L., House, C. H. & Brantley, S. L. (2005) Biogeochemical signatures through time as inferred from whole microbial genomes. *American Journal of Science*, **305**, 467–502.

2

The Architecture of Microbial Genomes

Introduction

Bacterial and archaeal genomes are organized and structured in several different ways. Forms of organization include physical structure, nucleotide composition, genes and regulatory regions, operons, and higher-order units such as replichores. Collectively, this organization of a genome can be referred to as genome architecture. It is influenced by a variety of biochemical, ecological, and evolutionary processes and forces. Understanding the architecture of microbial genomes and the processes that shape it is critical for understanding several aspects of community omic data.

2.1 Genome Size, Organization, and Replication

The genome architecture of bacteria and archaea is qualitatively similar, and distinct from that of eukarya (Koonin & Wolf 2008). At the highest level, bacterial and archaeal genomes are organized into chromosomes. Many bacteria and archaea have just one chromosome (Fig. 2.1), but others can have multiple chromosomes. The chromosome is generally more stable than plasmids, which are smaller self-replicating pieces of DNA. Plasmids are more dynamic, being able to merge, split, and rearrange frequently.

Double-stranded DNA is copied by replication in both directions ("bidirectional replication") from a single point of origin ("ori"). DNA is copied in the 5' to 3' direction, with a helicase unzipping the two strands at the replication fork. The strand being copied in the same direction as replication fork movement is copied continuously and is referred to as the leading strand. The other strand, being copied in the direction opposite replication

Genomic Approaches in Earth and Environmental Sciences, First Edition. Gregory Dick.
© 2019 John Wiley & Sons Ltd. Published 2019 by John Wiley & Sons Ltd.

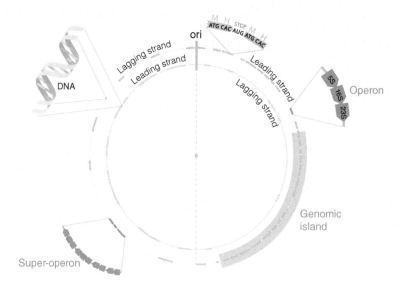

Figure 2.1 Genome organization in bacteria and archaea. Genomes typically consist of one circular chromosome (shown here), though multiple chromosomes and/or plasmids are possible. Each circle represents one of the two strands of DNA. Arrows represent genes and blocks of arrows are operons. See text for more details.

fork movement, must be synthesized in fragments ("Okazaki fragments") as the template becomes available. The leading and lagging strands are different in terms of nucleotide composition (see below), gene orientation, and gene content. Many essential, highly expressed genes are co-oriented with replication on the leading strand, which may be advantageous for minimizing collisions between replicating DNA polymerase/helicase and RNA polymerase (Koonin 2009). Also, because a new round of genome replication is often initiated before the previous round finishes, genes near the origin of replication can have up to eight times more copies per cell than genes near the terminus (Couturier & Rocha 2006). This effect is referred to as gene dose. Highly expressed genes are also often located near the origin of replication, in part to take advantage of the higher gene dose.

The genomes of bacteria and archaea are compact compared to those of eukaryotes, particularly multicellcular eukaryotes (Koonin 2009). Average coding density, i.e., the portion of a genome covered by genes, is 87%, and there is a strong correlation between the total number genes in a genome and the genome size (Kuo et al. 2009). However, there are exceptions. The cyanobacterial genus *Trichodesmium* has a coding percentage of ~60%, and many of the intergenic regions are expressed for unknown reasons (Walworth et al. 2015). Unlike eukaryotic genomes, genes in bacteria and archaea are generally not interrupted by introns, though there are exceptions to that rule as well (Simon et al. 2008; Tocchini-Valentini et al. 2011).

The size of bacterial and archaeal genomes varies across a 20-fold range, from 0.14 Mb to 13 Mb (Koonin & Wolf 2008; McCutcheon & Moran 2012). In fact, the lower range of bacterial and archaeal genome size overlaps with the upper range of viral genomes. Hence, there is no separation between the genome size of viral and cellular life (Koonin 2009). The lower range of cellular genomes is dominated by symbiotic bacteria; the smallest genome of a free-living organism is 580 kb (McCutcheon & Moran 2012). Genome size is not necessarily correlated with phylogeny (Bentley & Parkhill 2004), but rather is a function of the evolutionary forces that shape it, including both neutral and selective processes (see section 2.3).

A principal unit of organization of microbial genomes is the operon, a group of genes that are co-transcribed and thus regulated together. The intergenic space between genes within an operon (~0 bp) is shorter than that between operons (~100 bp) (Koonin & Wolf 2008). Operons are typically small, consisting of 2–4 genes, and are more strongly conserved in bacterial and archaeal genomes than is large-scale synteny (Koonin 2009). Operons often encode physically interacting proteins such as the various components of an ATP-dependent transporter (transmembrane, periplasmic, and ATPase subunits) (Koonin 2009). Such organization may help ensure correct balance of copy number of genes encoding subunits of protein complexes. Operons may be larger, such as those encoding 14 photosynthesis genes (Liotenberg et al. 2008) or over 50 ribosomal proteins (Koonin 2009). Such "super-operons" may be polycistronic (a single mRNA encodes multiple proteins), but there are also conserved groups of operons that form a "conserved gene neighborhood," or uberperon (Koonin 2009). Interestingly, the extent to which genomes are organized into operons varies widely; genomes like *Thermatoga maritima* are mostly organized into operons whereas some groups like cyanobacteria contain fewer operons (Koonin 2009). The factors that govern degree of "operonization" are not well understood, but it likely has to do with the balance of processes that disrupt operons (e.g., genome rearrangements) versus those that preserve them (e.g., horizontal gene transfer events that promote their survival and spread). Two main theories have been put forward to explain the existence of operons: (1) efficiency of regulation (Rocha 2008) and (2) the selfish gene/operon theory (Koonin 2009).

A major conclusion from comparative genomic studies is that gene order is usually poorly conserved beyond the operon scale for all but the most closely related strains. Comparison of gene order is effectively accomplished with dot plots, which often display patterns indicative of inversions around the origin of replication, suggesting that rearrangements that occur during replication are a major process affecting the organization of genomes (Eisen et al. 2000).

Genomes of closely related microorganisms share large regions of conserved genes. However, these regions are punctuated by hypervariable regions called genomic islands (also chromosomal islands). Genomic islands consist of 10–100 genes that are highly variable and dynamic. These regions often contain genes encoding specialized functions that are not needed for simple survival but which may encode important ecological adaptations that differentiate strains or populations. For example, genomic islands can include genes for biodegradation of anthropogenic compounds such as aromatic hydrocarbons and pesticides, genes encoding key aspects of pathogenesis (i.e., pathogenicity islands), and genes for the formation of magnetosomes, intracellular magnetite minerals that orient organisms with respect to Earth's magnetic field and aid in navigation through redox gradients. Genes encoding proteins that are used for viral attachment are also common in genomic islands (Avrani et al. 2011; Rodriguez-Valera et al. 2009).

The variable gene content within genomic islands leads to extraordinary diversity of gene content within microbial species. Genes that are variable across genomes at some taxonomic level (e.g., species) are referred to as flexible genes or accessory genes, whereas those that are present in all genomes are designated core genes (Cordero & Polz 2014). Significant shuffling of the core genome takes hundreds of millions of years whereas the flexible genome is much more dynamic (Rocha 2008). Taken together, the core and flexible genes represent the entire pool of genetic diversity of a species or higher taxonomic level, and this is referred to as the pangenome. The flexible genome also includes mobile elements such as plasmids, transposable elements, and integrated viruses (prophage). The degree of mobile elements in a genome varies tremendously and reflects the ecological and evolutionary characteristics of the microorganism. For example, the harmful algal bloom-forming cyanobacterium *Microcystis aeruginosa* contains an exceptional number of transposes, which it appears to use to rapidly rearrange the genome to a degree that may be regulated by nutrient availability (Steffen et al. 2014).

DNA sequences that are repeated within microbial genomes can occupy substantial portions of microbial genomes (Walworth et al. 2015). These elements can be important in the evolution of microbial genomes by mediating genome rearrangements (Darmon & Leach 2014). Intriguingly, such genome rearrangements may be regulated in response to environmental conditions, suggesting an evolutionary strategy for rapidly adapting to the environment (Steffen et al. 2014). These repeat sequences can wreak havoc on efforts to assemble microbial genomes and metagenomes (see sections 5.2 and 6.3).

2.2   Nucleotide Composition

The nucleotide composition of microbial genomes varies both within and across genomes. Within genomes, there is typically a difference in base composition between the leading and lagging strand. The leading strand is enriched in G and T whereas the lagging strand is enriched in A and C. This bias is often presented in terms of the percentage of excess of G over C (G-C/G+C) and is known as GC-skew. It can be easily visualized on a microbial genome and it marks the origin and terminus of replication (Rocha 2008). The cause of GC-skew is still not fully understood but likely relates to different mutational pressures that result from the different amount of time that the leading and lagging strands spend as single-stranded DNA, which is more prone to mutation (Rocha 2008).

GC content is another commonly used measure of nucleotide composition. It measures the percentage of nucleotides that are G and C (i.e., (G+C)/(A+T+C+G)). GC content varies widely across different microbial genomes, from approximately 25% to 85%, a fact that has been recognized since the early days of molecular biology (Sueoka 1962). In general, GC content is not correlated with taxonomy at higher levels, but some groups show cohesiveness in this regard, such as the high-GC gram positives (Actinobacteria) (Bentley & Parkhill 2004). The drivers of nucleotide composition include both neutral (Sueoka 1988) and selective forces (Foerstner et al. 2005; Rocha 2002) and their relative importance is still debated (Bentley & Parkhill 2004).

GC content is strongly correlated with genome size and lifestyle. AT-rich genomes are often small and occur in obligate symbionts and/or stable environments. GC-rich genomes tend to be large and in organisms that inhabit complex, highly variable environments. This has led to conclusions that GC content is an adaptive feature of the microbial genome that is selected on the basis of energetics/resources (Rocha 2002) or extreme environmental conditions (high temperature, salinity) in which the stronger bonds between G and C may be advantageous (Musto et al. 2004). However, convincing arguments have been made that nucleotide composition is the result of neutral processes such as mutational bias (Lynch 2007). Recent molecular studies support this contention (Wu et al. 2012). In addition to varying between genomes, GC content varies within genomes and is often used to detect genomic islands that may be of foreign origin.

Intimately connected to GC content is the differential utilization of codons within and between organisms. Each amino acid (except for methionine) can be encoded by multiple different codons (e.g., UUA, UUG, CUU, CUC, CUA, and CUG all encode leucine). The frequency at which each codon is used is highly variable. For example, in the cyanobacterium *Synechocystis* sp. PCC 6803 UUG is used in 30% of codons encoding leucine, whereas CUA is used only 9% of the time. This so-called "codon bias" reflects the tRNA content of a genome (Dong et al. 1996) and certain aspects of optimizing translation (Rocha 2004). The link between codon usage and translational efficiency may also be related to mRNAs stability (Boel et al. 2016). Indeed, highly expressed genes often display a distinct codon usage from others, leading to a within-genome codon bias. However, there are codon biases that are pervasive across whole microbial genomes, and these biases are distinct and distinguishable between microbial genomes. Again, neutral mutational bias may play a primary role in shaping such between-genome codon usage.

Whatever the forces that shape nucleotide composition are, there are strong signatures of nucleotide composition between genomes in terms of GC content and oligonucleotide frequency, and these signatures can be exploited for the purposes of metagenomic binning (see Chapter 7).

2.3 Ecological and Evolutionary Aspects of Microbial Genomes

Here we will briefly explore the ecological and evolutionary forces that shape genome architecture and generate genomic diversity. As we will see in the next section, the extent of this genotypic diversity in microbial populations and communities is incredible, especially in terms of gene content.

Genomes are shaped by a number of different processes, some of which are adaptive but many of which are neutral. Adaptive processes are those that are shaped by natural selection, providing enhanced fitness. This includes genome reduction or "streamlining" in which smaller genomes are favored to minimize the energy and resource burden of genome replication (Giovannoni et al. 2014), genetic modifications that change phenotype resulting in avoidance of predation (e.g., by grazers or phage), and mutations that lead to enhanced performance such as improved nutrient uptake, adaptation to light quality or intensity, etc. However, many characteristics of microbial genomes are random, have no discernible effect on fitness, and are apparently not under selective pressure. Thus, they are said to be neutral or nonadaptive. Mutation is essentially a stochastic process (point mutation, horizontal gene transfer, gene duplication, gene loss), as is recombination.

The effect on the fitness of organisms of these mutations can be positive, negative, or neutral. Preservation, loss, or even fixation (i.e., replacing all other variants at the same site in the genome) of new variants within a population can occur through random genetic drift (Kuo et al. 2009). Thus, a neutral or even slightly deleterious mutation can become fixed in a genome, and conversely, a slightly adaptive mutation can be lost through drift. Genetic changes that increase the complexity of genomes, such as gene duplication, are often neutral or slightly deleterious. They are retained only when selective pressure is weak. However, since the original copy of duplicated genes still serves the original function, the new genes are free to evolve new functions and thus provide raw material for evolutionary innovation. Thus, mutations that were initially neutral can be subjected to selection at a later time. The strength of selection is directly tied to population size, with genomes of smaller populations shaped more by drift and genomes of larger populations more by selection (Lynch 2007).

The relative importance of these different mechanisms for generating genetic diversity varies across the branches of the tree of life. For example, sexual recombination and gene duplication are prevalent in eukaryotes, whereas the gene repertoire of bacteria and archaea is greatly shaped by horizontal gene transfer. In eukaryotes, overall genotypic variation is largely at the level of different gene alleles, whereas in bacteria and archaea, genotypes are heavily influenced by gene content (Cordero & Polz 2014). These differences are due to differences in population size as well as fundamental biological properties such as mode of reproduction, genome architecture, and cell biology (Koonin 2009). The relative importance of neutral and adaptive processes in shaping the gene content of bacteria is still under debate. Recent work suggests that turnover in the content of accessory genes through gene gain, loss, or duplication is largely driven by neutral processes (Andreani et al. 2017; Wolf et al. 2017), while others conclude that it is largely adaptive (McInerney et al. 2017).

The interplay of selective and neutral processes is illustrated in the context of microbial genome size. Selective forces include genome streamlining, the idea that replicating the genome (and synthesizing the associated proteins) is costly in terms of energy and resources. Thus, in large populations that inhabit stable, oligotrophic environments, there may be strong selective pressure to maintain a small genome and discard unnecessary genes (Giovannoni et al. 2014). In the abundant and widespread marine bacterium SAR11, this results in small genomes with extreme coding density and reduced regulatory mechanisms (Giovannoni et al. 2005). The correlation of genome size with *rrn* copy number, resource availability, maximum reproductive rate, and growth efficiency (inversely) lends further support to an adaptive role in genome streamlining (Roller et al. 2016). Interestingly, genome reduction also occurs at the other end of the spectrum of effective populations size, where symbionts of eukaryotic hosts have extremely small genomes. This drastic genome reduction may be the result of processes that are fundamentally different from those acting during genome streamlining (Batut et al. 2014).

On the other hand, highly variable environments may select for organisms that are versatile, being able to compete under a range of different conditions. Such a versatile lifestyle requires not only more genes to encode more processes (e.g., different electron acceptors and donors for energy metabolism, more transporters and pathways for nutrient utilization), but also more genes for environmental sensing and regulation of gene expression. Indeed, with increasing genome size, an increasing proportion of the genome is dedicated to transcription and signal transduction (i.e., sensing and regulation) (Konstantinidis & Tiedje 2004).

Neutral processes, primarily genetic drift, can also exert strong influence on the evolution and size of microbial genomes (Lynch 2007). Gene duplication and acquisition of genes via horizontal gene transfer provide raw materials for genome evolution in the form of new genes. While pressures such as those above can selectively retain or purge such genes, in the absence of strong selection pressure the new genes may be retained even in the absence of any benefit to the organism. Hence, it is generally thought that neutral processes tend to enlarge genomes. However, Sela et al. (2016) recently provided support for a different theory – that gene acquisition is often beneficial, and that this is balanced against intrinsic bias for gene loss. A summary of ecological and evolutionary aspects of genome size is provided in Table 2.1.

Table 2.1 Ecological and evolutionary aspects of genome size.

Characteristic	Smaller genome	Larger genome
General ecology	Specialist	Generalist
	K strategist	R strategist*
Habitat stability	More stable	More variable
Environmental sensing and gene regulation	"Dumber"	"Smarter"
Population size	Large*	Small
Selection pressure	Strong	Weak

*Interestingly, in symbiotic bacteria the opposite is true: small population sizes result in small genomes (Batut et al. 2014). The dichotomy of R versus K strategies is almost certainly an oversimplification (Krause et al. 2014), and the linkage with genome size is debatable and likely more complicated than presented here.

Recent findings from different species of cyanobacteria illustrate the different evolutionary processes and patterns acting on microbial genomes. Evidence from hundreds of single cell genomes points to multiple distinct "genomic backbones" in *Prochlorococcus* (Kashtan et al. 2014). These are coherent and ancient core genomes that encode functionally distinct strains, or "ecotypes," of *Prochlorococcus*. In contrast, populations of the related *Synechococcus* have mosaic genomes in which recombination has apparently wiped away any vestiges of distinct genome-wide ecotypes (Rosen et al. 2015). This suggests that variable selection pressures act on many different individual loci across genomes, structuring diversity on local genomic scales (Desai & Walczak 2015; Retchless & Lawrence 2007). Understanding such dynamics is critical for defining genetically and ecologically cohesive units of microbial diversity (Cohan 2006; Shapiro & Polz 2014).

2.3.1   The Role of Viruses in Promoting Genomic Diversity

Viruses play an important role in promoting the genomic diversity of microorganisms. First, viruses gain access to host cells by attaching to specific proteins on the cell surface, leading to strong selection against these genes. In response, microbes diversify cell surface proteins in order to avoid viral predation. Interestingly, the genes encoding these highly variable proteins are clustered in hypervariable regions of the genome (Avrani et al. 2011) such as the genomic islands described earlier. Second, some viruses can integrate into the genomes of their hosts. Through this and other mechanisms, viruses can serve as vectors for horizontal gene transfer. The common occurrence of bacterial genes encoding metabolic processes such as photosynthesis (Mann et al. 2003) and sulfur oxidation (Anantharaman et al. 2014) suggests that viruses can play important roles in shaping the genetic underpinning of biogeochemical cycles.

2.4   Genomic Diversity in Microbial Communities

The form and extent of genomic diversity in a microbial community are important factors in considering strategies for analyzing and interpreting omics data from microbial communities, as well as potential pitfalls. Community genomic diversity occurs at several levels:

- diversity of microbial species (richness); different species will obviously have different genomic content
- microdiversity, i.e., clusters of closely related organisms containing less than 1% sequence divergence in the 16S rRNA gene (Acinas et al. 2004) can still have substantial genomic variability
- on an even finer scale, within-population diversity. For example, organisms that share identical 16S rRNA genes may still have substantially different genotypes due to single-nucleotide polymorphisms and variable gene content and order.

This section reviews these forms of genomic diversity as they occur in microbial communities, building on the previous section, which considered the ecological and evolutionary causes and effects of this diversity.

The richness of a microbial community is a key parameter in determining the number of genomes present in a community (see section 4.2.2) and hence the sequencing effort required to cover those genomes. Microbial community richness varies widely, from relatively simple communities in extreme environments such as acid mine drainage (Tyson et al. 2004) and subsurface fracture fluids (Chivian et al. 2008) (just a few to less than 100 species), to moderate-diversity communities such as seawater (hundreds to thousands of species), to high-diversity communities such as those found in soil and sediments (thousands of species) (Curtis et al. 2002; Gans et al. 2005; Howe et al. 2014). Just as crucial as richness is evenness; even in the early days of next-generation sequencing, genomes could be assembled from communities with high richness as long as the community had some dominant members (Lesniewski et al. 2012).

The diversity within species (e.g., within an operational taxonomic unit (OTU) defined at 97% sequence identity of the 16S rRNA gene) also exerts tremendous influence on the outcomes of metagenomic assemblies. Such intra-OTU diversity has been shown to be prevalent in natural microbial communities (Acinas et al. 2004). It is readily visualized through fragment recruitment of metagenomic data to microbial genomes (Rodriguez-R & Konstantinidis 2014). Depending on the degree of divergence and the assembler and parameters used, sequences from closely related populations may co-assemble into the same contigs/scaffolds (see sections 5.2 and 5.3 for definition and examples) or into separate contigs/scaffolds (Allen & Banfield 2005). Variation in gene order and content between populations, which is prevalent even between closely related genotypes (Koonin 2009), will result in the premature termination of contigs/scaffolds in those cases where distinct populations co-assemble. Even in cases where populations generally assemble into separate contigs/scaffolds, highly conserved genes or repeat regions (e.g., transposons) may assemble together, potentially confounding the assembly process.

This type of intra-OTU genomic diversity may help to explain how different OTUs in the same community can produce very different assembly outcomes. For example, in microbial communities in hydrothermal plumes of the Guaymas Basin, SUP05, SAR324, and MGI Archaea genomes were effectively assembled whereas those of *Methylococcacea* were not, even though the latter were present at greater abundance at the OTU level (Lesniewski et al. 2012). The lack of clonality, or "genomic coherence," has been observed to complicate assembly efforts in even low-diversity communities that are dominated by just a few species (Teeling & Glockner 2012).

Genomic diversity also occurs at fine phylogenetic scales, within populations. Allen et al. (2007) resolved genomic differences between two closely related genomes of *Ferroplasma acidarmanus*, one from an isolate and one an environmental population at the same site. Although these genotypes shared 99.8% nucleotide similarity of 16S rRNA genes, considerable differences in gene content and order were resolved. Much of this genomic variation was ascribed to rearrangements due to transposable elements and integration of prophage into the genomes. The major role of transposable elements in rearranging genomes has also been demonstrated for strains of the toxin-producing cyanobacterium *Microcystis aeruginosa* (Humbert et al. 2013). Intriguingly, transcription of these transposons is regulated by nutrient availability, providing a window into how the evolutionary trajectory of microbial genomes can be directly influenced by the environment, and potentially by anthropogenic activity (Steffen et al. 2014). Genomic divergence between close relatives has been associated with ecological divergence (Denef & Banfield 2012; Denef et al. 2010; Lo et al. 2007; Pena et al. 2010). Phage also play a major role in population level heterogenetity (Avrani et al. 2011; Tyson & Banfield 2008).

Because of these multiple dimensions of genomic variation, there is incredible diversity of genes in natural microbial communities. At this stage of technological development, this diversity appears to be essentially limitless, and metagenomic sequencing has vastly expanded the known genetic and biochemical diversity of life on Earth (Yooseph et al. 2007). Even in microbial communities for which there are dozens of representative cultures with genome sequences, a vast portion of the metagenomic reads does not map to the reference genomes (Oh et al. 2014). Because of the typical architecture of microbial genomes, with large blocks of conserved/syntenous genes interrupted by regions of highly variable gene content (genomic islands – see section 2.1), the core and flexible portions of genomes can be expected to exhibit different outcomes of assembly in terms of coverage, contig length, and sequence polymorphism.

2.5   Does Genomic Diversity Matter?

To summarize the discussion in the above sections, the evolution of microbial genomes can be viewed as a balance between neutral processes that create disorder and selective processes that can organize genomes (Koonin 2009). As a result, microbial genomes are of course far from random, but they are also far from having optimal architecture for biological functioning (Koonin 2009). Although it is generally recognized that both neutral and selective processes shape microbial genomes, their relative importance is debated. This debate extends to the question of whether the extensive genomic diversity in microbial communities "matters." In other words, are the many novel genes of unknown function that are commonly observed in microbial genomes, and especially metagenomic datasets, important ecologically? Or do they simply reflect neutral processes such as horizontal transfer of genes that are of little or no consequence to the organisms and communities that host them?

Several considerations suggest that much of the observed microbial genomic diversity is neutral. First is the extent of genomic diversity. Even strains that are nearly identical in sequence across the core genome can have hundreds of unique genes, and there are thousands of rare genotypes within strains (Thompson et al. 2005). Second, there are considerations relating to the relatively dilute nature of microbial genotypes within natural habitats that suggest that strong competitive interactions between genotypes are uncommon in time and space (Cordero & Polz 2014).

On the other hand, several lines of evidence indicate that the genomic diversity in microbes is tremendously important. In some cases, the content of the flexible genome correlates with phylogeny and encodes important interactions with the local environment that differentiate the ecology of closely related organisms (Kashtan et al. 2014). Indeed, the flexible genome is biased in terms of functions, often encoding proteins for the cell surface proteins, DNA binding, and pathogenesis (Nakamura et al. 2004). Further supporting their ecological importance, genes in the flexible genome are often highly expressed in the environment (Frias-Lopez et al. 2008; Ram et al. 2005). However, it should be noted that the portion of flexible genes that are ecologically important may be small, as suggested by observations that the vast majority (>85%) of predicted proteins in genomes of acid mine drainage microorganisms were never identified in proteomics datasets across 27 samples (Denef et al. 2010).

References

Acinas, S. G., Klepac-Ceraj, V., Hunt, D. E., et al. (2004) Fine-scale phylogenetic architecture of a complex bacterial community. *Nature*, **430**, 551–554.

Allen, E. E. & Banfield, J. F. (2005) Community genomics in microbial ecology and evolution. *Nature Reviews Microbiology*, **3**, 489–498.

Allen, E. E., Tyson, G. W., Whitaker, R. J., Detter, J. C., Richardson, P. M. & Banfield, J. F. (2007) Genome dynamics in a natural archaeal population. *Proceedings of the National Academy of Sciences of the United States of America* **104**, 1883–1888.

Anantharaman, K., Duhaime, M. B., Breier, J. A., Wendt, K. A., Toner, B. M. & Dick, G. J. (2014) Sulfur oxidation genes in diverse deep-sea viruses. *Science*, **344**, 757–760.

Andreani, N. A., Hesse, E. & Vos, M. (2017) Prokaryote genome fluidity is dependent on effective population size. *ISME Journal*, **11**, 1719–1721.

Avrani, S., Wurtzel, O., Sharon, I., Sorek, R. & Lindell, D. (2011) Genomic island variability facilitates Prochlorococcus-virus coexistence. *Nature*, **474**, 604–608.

Batut, B., Knibbe, C., Marais, G. & Daubin, V. (2014) Reductive genome evolution at both ends of the bacterial population size spectrum. *Nature Reviews Microbiology*, **12**, 841–850.

Bentley, S. D. & Parkhill, J. (2004) Comparative genomic structure of prokaryotes. *Annual Review of Genetics*, **38**, 771–792.

Boel, G., Letso, R., Neely, H., et al. (2016) Codon influence on protein expression in E. coli correlates with mRNA levels. *Nature*, **529**, 358.

Chivian, D., Brodie, E. L., Alm, E. J., et al. (2008) Environmental genomics reveals a single–species ecosystem deep within Earth. *Science*, **322**, 275–278.

Cohan, F. M. (2006) Towards a conceptual and operational union of bacterial systematics, ecology, and evolution. *Philosophical Transactions of the Royal Society B – Biological Sciences*, **361**, 1985–1996.

Cordero, O. X. & Polz, M. F. (2014) Explaining microbial genomic diversity in light of evolutionary ecology. *Nature Reviews Microbiology*, **12**, 263–273.

Couturier, E. & Rocha, E. P. C. (2006) Replication-associated gene dosage effects shape the genomes of fast-growing bacteria but only for transcription and translation genes. *Molecular Microbiology*, **59**, 1506–1518.

Curtis, T. P., Sloan, W. T. & Scannell, J. W. (2002) Estimating prokaryotic diversity and its limits. *Proceedings of the National Academy of Sciences of the United States of America*, **99**, 10494–10499.

Darmon, E. & Leach, D. R. F. (2014) Bacterial genome instability. *Microbiology and Molecular Biology Reviews*, **78**, 1–39.

Denef, V. J. & Banfield, J. F. (2012) In situ evolutionary rate measurements show ecological success of recently emerged bacterial hybrids. *Science*, **336**, 462–466.

Denef, V. J., Kalnejais, L. H., Mueller, R. S., et al. (2010) Proteogenomic basis for ecological divergence of closely related bacteria in natural acidophilic microbial communities. *Proceedings of the National Academy of Sciences of the United States of America*, **107**, 2383–2390.

Desai, M. M. & Walczak, A. M. (2015) Flexible gene pools. *Science*, **348**, 977–978.

Dong, H. J., Nilsson, L. & Kurland, C. G. (1996) Co-variation of tRNA abundance and codon usage in Escherichia coli at different growth rates. *Journal of Molecular Biology*, **260**, 649–663.

Eisen, J. A., Heidelberg, J. F., White, O., Salzberg, S. L. (2000) Evidence for symmetric chromosomal inversions around the replication origin in bacteria. *Genome Biology*, **1**, research0011.1–0011.9.

Foerstner, K. U., von Mering, C., Hooper, S. D. & Bork, P. (2005) Environments shape the nucleotide composition of genomes. *EMBO Reports*, **6**, 1208–1213.

Frias-Lopez, J., Shi, Y., Tyson, G. W., et al. (2008) Microbial community gene expression in ocean surface waters. *Proceedings of the National Academy of Sciences of the United States of America* **105**, 3805–3810.

Gans, J., Wolinsky, M. & Dunbar, J. (2005) Computational improvements reveal great bacterial diversity and high metal toxicity in soil. *Science*, **309**, 1387–1390.

Giovannoni, S. J., Tripp, H. J., Givan, S., et al. (2005) Genome streamlining in a cosmopolitan oceanic bacterium. *Science*, **309**, 1242–1245.

Giovannoni, S. J., Thrash, J. C. & Temperton, B. (2014) Implications of streamlining theory for microbial ecology. *ISME Journal*, **8**, 1553–1565.

Howe, A. C., Jansson, J. K., Malfatti, S. A., Tringe, S. G., Tiedje, J. M. & Brown, C. T. (2014) Tackling soil diversity with the assembly of large, complex metagenomes. *Proceedings of the National Academy of Sciences of the United States of America*, **111**, 6115–6115.

Humbert, J. F., Barbe, V., Latifi, A., et al. (2013) A tribute to disorder in the genome of the bloom-forming freshwater cyanobacterium Microcystis aeruginosa. *Plos One*, **8**, e70747.

Kashtan, N., Roggensack, S. E., Rodrigue, S., et al. (2014) Single-cell genomics reveals hundreds of coexisting subpopulations in wild Prochlorococcus. *Science*, **344**, 416–420.

Konstantinidis, K. T. & Tiedje, J. M. (2004) Trends between gene content and genome size in prokaryotic species with larger genomes. *Proceedings of the National Academy of Sciences of the United States of America*, **101**, 3160–3165.

Koonin, E. V. (2009. Evolution of genome architecture. *International Journal of Biochemistry and Cell Biology*, **41**, 298–306.

Koonin, E. V. & Wolf, Y. I. (2008) Genomics of bacteria and archaea: the emerging dynamic view of the prokaryotic world. *Nucleic Acids Research*, **36**, 6688–6719.

Krause, S., Le Roux, X., Niklaus, P. A., et al. (2014) Trait-based approaches for understanding microbial biodiversity and ecosystem functioning. *Frontiers in Microbiology*, **5**, 251.

Kuo, C. H., Moran, N. A. & Ochman, H. (2009) The consequences of genetic drift for bacterial genome complexity. *Genome Research*, **19**, 1450–1454.

Lesniewski, R. A., Jain, S., Anantharaman, K., Schloss, P. D. & Dick, G. J. (2012) The metatranscriptome of a deep-sea hydrothermal plume is dominated by water column methanotrophs and lithotrophs. *ISME Journal*, **6**, 2257–2268.

Liotenberg, S., Steunou, A. S., Picaud, M., Reiss-Husson, F., Astier, C. & Ouchane, S. (2008) Organization and expression of photosynthesis genes and operons in anoxygenic photosynthetic proteobacteria. *Environmental Microbiology*, **10**, 2267–2276.

Lo, I., Denef, V. J., Verberkmoes, N. C., et al. (2007) Strain-resolved community proteomics reveals recombining genomes of acidophilic bacteria. *Nature*, **446**, 537–541.

Lynch, M. (2007) *The Origins of Genome Architecture*. Sinauer Associates, Inc., Sunderland, MA.

Mann, N. H., Cook, A., Millard, A., Bailey, S. & Clokie, M. (2003) Marine ecosystems: bacterial photosynthesis genes in a virus. *Nature*, **424**, 741.

McCutcheon, J. P. & Moran, N. A. (2012) Extreme genome reduction in symbiotic bacteria. *Nature Reviews Microbiology*, **10**, 13–26.

McInerney, J. O., McNally, A. & O'Connell, M. J. (2017) Why prokaryotes have pangenomes. *Nature Microbiology*, **2**, 17040.

Musto, H., Naya, H., Zavala, A., Romero, H., Alvarez-Valin, F. & Bernardi, G. (2004) Correlations between genomic GC levels and optimal growth temperatures in prokaryotes. *Febs Letters*, **573**, 73–77.

Nakamura, Y., Itoh, T., Matsuda, H. & Gojobori, T. (2004) Biased biological functions of horizontally transferred genes in prokaryotic genomes. *Nature Genetics*, **36**, 760–766.

Oh, J., Byrd, A. L., Deming, C., et al. (2014) Biogeography and individuality shape function in the human skin metagenome. *Nature*, **514**, 59.

Pena, A., Teeling, H., Huerta-Cepas, J., et al. (2010) Fine-scale evolution: genomic, phenotypic and ecological differentiation in two coexisting Salinibacter ruber strains. *ISME Journal*, **4**, 882–895.

Ram, R. J., Verberkmoes, C., Thelen, M. P., et al. (2005) Community Proteomics of a natural microbial biofilm. *Science*, **308**, 1915–1920.

Retchless, A. C. & Lawrence, J. G. (2007) Temporal fragmentation of speciation in bacteria. *Science*, **317**, 1093–1096.

Rocha, E. P. C. (2002) Base composition might result from competition for metabolic resources. *Trends in Genetics*, **18**, 291–294.

Rocha, E. P. (2004) Codon usage bias from tRNA's point of view: redundancy, specialization, and efficient decoding for translation optimization. *Genome Research*, **14**, 2279–2286.

Rocha, E. P. C. (2008) The organization of the bacterial genome. *Annual Review of Genetics*, **42**, 211–233.

Rodriguez-R, L. M. & Konstantinidis, K. T. (2014) Bypassing cultivation to identify bacterial species. *Microbe*, **9**, 111–118.

Rodriguez-Valera, F., Martin-Cuadrado, A. B., Rodriguez-Brito, B., et al. (2009) OPINION: Explaining microbial population genomics through phage predation. *Nature Reviews Microbiology*, **7**, 828–836.

Roller, B. R., Stoddard, S. F. & Schmidt, T. M. (2016) Exploiting rRNA operon copy number to investigate bacterial reproductive strategies. *Nature Microbiology*, **1**, 16160.

Rosen, M. J., Davison, M., Bhaya, D. & Fisher, D. S. (2015) Fine-scale diversity and extensive recombination in a quasisexual bacterial population occupying a broad niche. *Science*, **348**, 1019–1023.

Sela, I., Wolf, Y. I. & Koonin, E. V. (2016) Theory of prokaryotic genome evolution. *Proceedings of the National Academy of Sciences of the United States of America*, **113**, 11399–11407.

Shapiro, B. J. & Polz, M. F. (2014) Ordering microbial diversity into ecologically and genetically cohesive units. *Trends in Microbiology*, **22**, 235–247.

Simon, D. M., Clarke, N. A. C., McNeil, B. A., et al. (2008) Group II introns in eubacteria and archaea: ORF-less introns and new varieties. *RNA*, **14**, 1704–1713.

Steffen, M. M., Dearth, S. P., Dill, B. D., et al. (2014) Nutrients drive transcriptional changes that maintain metabolic homeostasis but alter genome architecture in Microcystis. *ISME Journal*, **8**, 2080–2092.

Sueoka, N. (1962) On the genetic basis of variation and heterogeneity of DNA base composition. *Proceedings of the National Academy of Sciences of the United States of America*, **48**, 582–592.

Sueoka, N. (1988) Directional mutation pressure and neutral molecular evolution. *Proceedings of the National Academy of Sciences of the United States of America*, **85**, 2653–2657.

Teeling, H. & Glockner, F. O. (2012) Current opportunities and challenges in microbial metagenome analysis – a bioinformatic perspective. *Briefings in Bioinformatics*, **13**, 728–742.

Thompson, J. R., Pacocha, S., Pharino, C., et al. (2005) Genotypic diversity within a natural coastal bacterioplankton population. *Science*, **307**, 1311–1313.

Tocchini-Valentini, G. D., Fruscoloni, P. & Tocchini-Valentini, G. P. (2011) Evolution of introns in the archaeal world. *Proceedings of the National Academy of Sciences of the United States of America*, **108**, 4782–4787.

Tyson, G. W. & Banfield, J. F. (2008) Rapidly evolving CRISPRs implicated in acquired resistance of microorganisms to viruses. *Environmental Microbiology*, **10**, 200–207.

Tyson, G. W., Chapman, J., Hugenholtz, P., et al. (2004) Community structure and metabolism through reconstruction of microbial genomes from the environment. *Nature*, **428**, 37–43.

Walworth, N., Pfreundt, U., Nelson, W. C., et al. (2015) Trichodesmium genome maintains abundant, widespread noncoding DNA in situ, despite oligotrophic lifestyle. *Proceedings of the National Academy of Sciences of the United States of America*, **112**, 4251–4256.

Wolf, Y. I., Makarova, K. S., Lobkovsky, A. E. & Koonin, E. V. (2017) Two fundamentally different classes of microbial genes. *Nature Microbiology*, **2**, 16208.

Wu, H., Zhang, Z., Hu, S. N. & Yu, J. (2012) On the molecular mechanism of GC content variation among eubacterial genomes. *Biology Direct*, **7**, 2.

Yooseph, S., Sutton, G., Rusch, D. B., et al. (2007) The *Sorcerer II* Global Ocean Sampling expedition: expanding the universe of protein families. *PLoS Biology*, **5**, e16.

3

Application of Omics Approaches to Earth and Environmental Sciences

Opportunities and Challenges

Introduction

Now armed with a basic understanding of microbial genomics and an appreciation of the short but eventful history leading up to the current point in time, we are prepared to consider the potential for omics approaches to advance the Earth and environmental sciences. The applications are diverse. Omics approaches can help trace geochemical processes and intermediates that are difficult to detect with chemical approaches; reveal how biogeochemical cycles are coupled; provide new dimensions of data for building and evaluating biogeochemical and climate models; show how biomarkers are distributed across the tree of life; and serve as a valuable complement to the rock record in tracing the history of life and its co-evolution with geochemistry. Portions of this text were drawn from early drafts of Dick and Lam (2015).

3.1 New Perspectives on Microbial Biogeochemistry

3.1.1 Redefining the Carbon and Nitrogen Cycles

Omics approaches provide entirely new ways of determining which organisms are responsible for specific biogeochemical processes and for revealing the biochemical mechanisms that underpin these processes. One of the first major advances using this approach was the discovery of previously unknown mechanisms of harnessing sunlight for energy and fixation of carbon. This included the recognition that light-driven proton pumps (Béjà et al. 2000) and nonoxygen-producing photosynthesis (Béjà et al. 2002) are

Genomic Approaches in Earth and Environmental Sciences, First Edition. Gregory Dick.
© 2019 John Wiley & Sons Ltd. Published 2019 by John Wiley & Sons Ltd.

significant pathways of phototrophy in the world's oceans (Karl 2002). Early environmental genomics studies also played a key role in deciphering the pathway for anaerobic methane oxidation (Hallam et al. 2004; Pernthaler et al. 2008). More recently, proteomics revealed patterns of nutrient stress in marine cyanobacteria, providing new insights into the factors that limit primary production and hence deepening our understanding of the carbon cycle (Saito et al. 2014). Omics approaches also played a key role in revealing unexpectedly high potential for dark carbon fixation based on lithotrophy in the deep ocean (Anantharaman et al. 2016a; Aristegui et al. 2009; Reed et al. 2015; Swan et al. 2011), the role of viral auxiliary metabolic genes in the carbon cycle (Thompson et al. 2011), and the existence of diverse microbial groups, including candidate phyla with no cultured representatives (Brown et al. 2015) that appear to play key roles in the biogeochemical cycling of carbon and other elements (Solden et al. 2016; Wrighton et al. 2014).

Omics approaches also revolutionized our understanding of the nitrogen cycle. Early on, the discovery of ammonia monooxygenase on a fragment of archaeal DNA led to the realization that dominant marine archaea (Venter et al. 2004) are involved in ammonia oxidation (Francis et al. 2005; Könneke et al. 2005). Omics approaches were also key to elucidating the evolutionary history and biochemical pathways of anaerobic ammonia oxidation (Strous et al. 2006). This work enabled the development of genetic probes to track these processes in the environment, providing new data streams on the abundance and activity of different functional groups and new means of resolving major biogeochemical issues such as the pathways responsible for nitrogen loss in the oceans (Kraft et al. 2014; Lam et al. 2009; Ward et al. 2009). In an elegant demonstration of the complementary powers of omics and physiology, Ettwig et al. (2010) uncovered the intricate mechanisms of how methane oxidation is coupled to nitrite reduction via an O_2 intermediate.

Where essentially complete microbial genomes can be reconstructed directly from environmental metagenomic data, the coupling of biogeochemical cycles can be revealed, as in the case of the sulfur and nitrogen cycles in oxygen minimum zones (Walsh et al. 2009). Just when the potential for major game-changing advances in understanding the nitrogen cycle seemed to be dwindling, we encounter two more big surprises. In 2015, several different research groups independently used genome assembly to discover that ammonia oxidation and nitrite oxidation, processes long thought to be split into two different organisms, can in fact be housed in the same organism (Daims et al. 2015, 2016; Pinto et al. 2015; van Kessel et al. 2015). Then, in 2016, single cell genomics and transcriptomics were used to show that certain lineages of SAR11, one of the most abundant groups of organisms in the oceans, are adapted to oxygen minimum zones, where they denitrify nitrate to N_2, thus participating in the main pathway for loss of oceanic N_2 (Tsementzi et al. 2016).

3.1.2   Omics as Sensitive and Efficient Tracers of Biogeochemical Processes

Omics approaches can also serve as a uniquely sensitive tracer of biogeochemistry. Whereas transient chemical intermediates may be difficult to detect by geochemical approaches, genes, transcripts, or proteins that are signatures of those chemical intermediates may be readily apparent. For example, Canfield et al. (2010) uncovered a cryptic sulfur cycle in which oxidation of organic carbon fuels bacterial sulfate reduction, producing sulfide. The cycle is cryptic in that the sulfide does not accumulate because it is immediately oxidized back to sulfate by sulfide-oxidizing bacteria. In such cases where geochemical cycling proceeds rapidly and without the build-up of products, metagenomic data can serve as a form of biogeochemical reconnaissance that generates hypotheses, which can then be tested with targeted methods as Canfield and colleagues did with isotopic labeling approaches.

More broadly than the value omics data affords for detecting cryptic processes, metagenomics can serve as a valuable tracer of biogeochemistry simply due to the sheer scale and geographic breadth of data in public databases, which are growing at an ever-increasing rate. For example, Podar et al. (2015) recently leveraged 3500 publicly available microbial genomes to trace the global distribution of genes for mercury methylation in nature. Hence, as long as datasets are preserved in an accessible and usable form, omics data can serve as a complement or even a forerunner to geochemical measurements for understanding the biogeography of biogeochemical processes.

3.1.3   Omics Data is Valuable for Biogeochemical Models

As omics data from environments around the world accumulates at an accelerating pace, this information represents a valuable new resource for the development and evaluation of biogeochemical models. Although gaps in knowledge regarding the links between gene sequences and biochemical traits (e.g., enzymatic function, substrate specificity, kinetics) present a formidable challenge (see section 3.3), initial efforts to incorporate omics data into biogeochemical models show promising results (Reed et al. 2014). Incorporation of omics data into models is discussed in detail in Chapter 10.

3.1.4   Understanding Biotic Responses and Feedbacks to Global Change

Because microorganisms are a critical component in the response of the biosphere to global change, omics approaches can provide key insights into the biological feedbacks of global change on biogeochemical cycles and climate.

For example, microorganisms will likely influence the fate of carbon in permafrost, which holds roughly as much organic carbon as that in land plants and the atmosphere. Using a metagenomic approach, Mackelprang and colleagues showed that microbial communities rapidly shift as permafrost thaws (Mackelprang et al. 2011). Methane is released during this thaw (both from previous accumulation and from new production via methanogenesis), but it is also rapidly consumed by methane-oxidizing organisms. The omics data in this study suggest that microorganisms generate and consume methane, thus revealing pathways that influence the flux of greenhouse gases emitted from thawing permafrost. As is often the case, the omics data generated a hypothesis which could then be tested with follow-up experimentation and/or quantitative models such as those described above.

Omics data can also be used to monitor microorganisms that conduct bioremediation and biodegradation, such as in revealing microorganisms involved in oil degradation following the Deepwater Horizon oil spill (Mason et al. 2012; Rivers et al. 2013). These opportunities have been perhaps demonstrated best in the sediment-hosted perennially suboxic/anoxic aquifer adjacent to the Colorado River, near Rifle, CO, USA, where omics approaches have been integrated with rich datasets from experiments and geochemistry. Numerous essentially complete genomes were recovered from complex communities in a uranium-contaminated aquifer where *in situ* acetate amendment was used to stimulate microbial reduction of soluble U(VI) to insoluble U(IV) (Wrighton et al. 2012). The metabolic status of these microbial communities was tracked with genomic and proteomic methods, yielding insights into the flux of energy and carbon and the rate-limiting steps of the bioremediation process (Wilkins et al. 2009).

Importantly, the omics approaches were conducted in a community context so that the flow of carbon, sulfur, and metals through the ecosystem could be tracked both in the context of the different organisms catalyzing each step and in the context of interlinking biogeochemical cycles (Anantharaman et al. 2016b; Wrighton et al. 2014).

3.2 A Genomic Record of Biological and Geochemical Evolution

Genome sequences record valuable information on the history of life and its evolution in the context of evolving geochemistry (Raymond, 2005; Zerkle et al. 2005). They encode the microorganism's potential to synthesize organic molecules, which may be preserved in the rock record as molecular fossils, or "biomarkers" (see Chapter 9). These molecules are often lipids, due to their exceptional preservation, and if they are specific to certain microbial groups or metabolisms then they can be used to infer the presence of those microbes and/or metabolisms in ancient ecosystems. Such biomarkers can also provide views into past environmental conditions (e.g., temperature, salinity, redox state).

How can we determine that certain biomarker molecules are only associated with certain organisms or metabolisms, and therefore represent a faithful signature of those organisms/metabolisms in the rock record? Traditionally, this has been done by screening microbial cultures for compounds of interest. However, because only a tiny fraction of microbial life has been brought into pure culture, lab cultures provide an incomplete picture of which microbes produce which compounds. Another problem is that some compounds may be produced only under certain conditions, so the absence of a compound in a culture does not mean the organism is incapable of making it.

Genomic perspectives have already demonstrated their value through studies of biomarkers for oxygenic photosynthesis. 2-Methylhopanes were long considered a biomarker of cyanobacteria, the organisms responsible for the oxygenation of Earth's surface, and these molecules were key evidence of the appearance of oxygenic photosynthesis on the early Earth (Summons et al. 1999). However, the genes putatively involved in biosynthesis of 2-methylhopanes were recently found to be present in the genomes of many noncyanobacteria that do not perform oxygenic photosynthesis (Welander et al. 2010). These results suggested that 2-methylhopanes should not be used as direct biomarkers of oxygenic photosynthesis. Conversely, genome sequences can be used to predict biosynthetic pathways and determine which compounds *are* unique to certain taxonomic groups or metabolisms. Combined with additional approaches and applied to natural communities of microorganisms, this could provide a map of how biomarkers are distributed across the tree of life (Brocks & Banfield 2009; Pearson 2014).

A second form of information recorded within genomes that can help trace Earth's geochemical evolution involves the function and metal content of predicted proteins. For example, gene families requiring the use of molecular oxygen (O_2), and associated metabolic networks, are expected to have expanded upon the widespread oxygenation of Earth's atmosphere (David & Alm, 2011; Raymond & Segre 2006). Similarly, the use of metals (e.g., Fe, Zn, Mn) as co-factors in proteins that participate in redox reactions may reflect the environmental availability of those metals, which has changed through Earth history along with changing ocean chemistry (Dupont et al. 2010; Glass 2015; Saito et al. 2003). Thus, analysis of modern genomes can help us to infer the order and patterns of evolutionary innovation as they relate to geochemistry. Further, models of the evolutionary history of proteins can shed light on the evolution of Earth's redox status and geochemistry (David and Alm, 2011; Rothman et al. 2014; Zerkle et al. 2005).

Finally, sequence information can provide insights into the timing of the evolution of major microbial metabolisms and lineages. For example, molecular phylogenetic data has been used to constrain the timing and pathways of the evolution of methanogenesis, phototrophy, cell morphology, nitrogen fixation, silica biomineralization, and colonization of land (Battistuzzi et al. 2004; Blank & Sanchez-Baracaldo 2010; Boyd & Peters 2013; Boyd et al. 2011; Schirrmeister et al. 2013; Trembath-Reichert et al. 2015). There are, however, difficulties in using molecular clocks to translate sequence information into actual dates (Pulquerio & Nichols 2007).

3.3   Challenges and Limitations of Omics Approaches

A practical issue that presents very real barriers to many researchers is the computational challenge of dealing with enormous datasets. This includes the first-order issues of having the necessary hardware to perform basic tasks of storage and processing. In contrast to the early days of DNA sequencing, the computational costs associated with storing and analyzing data now exceed those of producing the data! Next, there are limitations in software tools and expertise needed to accomplish tasks that geobiologists want to accomplish (see section 12.4).

There is also the challenge of data dissemination and access, which depends on issues of databases and their integration, standards for metadata and quality control, and data curation (Brown & Tiedje 2011; Gilbert et al. 2014). Large centers and pipelines such as IMG (Markowitz et al. 2009) and MG-RAST (Meyer et al. 2008) provide critical resources in terms of both analysis and databases, but have limitations in terms of turnaround times and breadth of analyses offered, and are typically not flexible to accommodate special user needs. CAMERA (Seshadri et al. 2007) was another "top-down" resource that served some users from the marine microbiology community well but was not responsive enough to many user needs. These challenges will only multiply as the field moves towards more comparative and experimental studies that involve large numbers of samples and utilize multidimensional forms of data (e.g., genes, bins, transcripts, strains across space and time), omics approaches, and parallel geochemical and environmental data.

An exciting trend in the omics field is to enable dissemination and support of user-developed applications through collaborative and open-source platforms (see section 12.4). Here, the plant biology community provides a model with the iPlant Collaborative, which grew into CyVerse (www.cyverse.org/) and was used as a foundation for iVirus (Bolduc et al. 2017). The US Department of Energy has also developed an open platform called kbase (https://kbase.us).

In addition to the practical issues discussed above, a major challenge inherent to omics data is lack of knowledge about the physiological or geochemical function of many genes and proteins. One of the most astounding insights to emerge from the omics age is that microbial life harbors incredible genetic diversity. This was recognized in early microbial genome sequencing efforts, where a large portion of apparent genes (predicted to be genes on the basis of "open reading frames," i.e., long stretches without a stop codon) were found to be novel and of unknown function (Roberts et al. 2004). Metagenomic studies have further revealed extraordinary, seemingly infinite genetic and biochemical diversity in natural microbial communities (Temperton & Giovannoni 2012; Yooseph et al. 2007). Genetic novelty is especially prevalent in the still numerous uncharted branches of the tree of life, so-called "microbial dark matter" (Rinke et al. 2013). An important question is: are such genes functionally important to their host organisms? For many genes, the answer is a resounding "yes." Laboratory experiments on model organisms show that genes encoding protein domains of unknown function are often biologically essential (Goodacre et al. 2014).

The preponderance of unknown microbial genes underscores the urgent need to step up genetic and biochemical studies conducted on model organisms or enrichment cultures. These laboratory experiments are a key avenue to uncovering relationships between genes, geochemistry, physiology, and novel metabolic pathways; there is no substitute for such "traditional" approaches and a renewed commitment to these methods is required (Newman et al. 2012). However, given the vast diversity of genes in the environment, higher-throughput methods are also needed to formulate hypotheses of links between genes and geochemistry and to prioritize genes and organisms for targeted studies. To some extent, this can be accomplished by developing high-throughput methods of genetic screens in model organisms (Deutschbauer et al. 2011). A powerful yet challenging approach is "functional metagenomics" in which environmental DNA is cloned into model laboratory organisms, which express the encoded proteins, which can then be screened functionally (Handelsman 2004; Taupp et al. 2011; Wrighton et al. 2016). Though this approach has been around from the beginning of genomics and has demonstrated its value in some cases, there are many challenges associated with expressing genes from an unknown organism in a lab culture ("heterologous expression"). Likewise, high-throughput expression screening of metagenomic libraries (Uchiyama et al. 2005) offers tantalizing potential but has not yet found widespread application. Renewed efforts to develop advanced technologies for functional annotation are much needed (Baric et al. 2016).

3.4 Omics as a Complement to Other Approaches

Given the inherent limitations of omics, how are these approaches best employed? First of all, it should be recognized that omics data represents a powerful resource for exploring microbial communities and *generating* hypotheses. This approach embraces the notion that "listening to what the microbes have to say" can provide incisive insights into key yet potentially unexpected processes such as cryptic sulfur cycling (Canfield et al. 2010; Paez-Espino et al. 2016). Omics tools are unparalleled in terms of hearing the language of microbes as it is whispered within natural microbial communities (i.e., detecting the presence and expression of metabolic pathways, nutrient status, cell–cell communications, etc.) (Moran 2009). The extraordinary value of the exploratory, hypothesis-generating capabilities of omics data has been thoroughly demonstrated (Jansson 2013).

Whereas omics is often viewed as an alternative to cultivation, these two approaches can also be seen as partners. For example, omics data can provide key information for cultivation strategies and direct biochemical studies to elucidate the enzymes that underpin geochemical reactions (Ram et al. 2005; Tyson et al. 2005). Indeed, omics approaches are most valuable when conducted in parallel with traditional geochemical and microbiological approaches in an integrated fashion (Oremland et al. 2005). An outstanding example of how omics can be leveraged as one of a suite of methods to probe a geomicrobial system is provided by Wilbanks et al. (2014). Gene surveys, omics, and microscopy were used to generate hypotheses, which were then tested with voltammetry and stable isotope approaches. In another example, Pernthaler et al. demonstrate the power of using cell capture to select certain microbial groups for omics analyses, yielding genomic predictions that methanotrophs are capable of fixing nitrogen. This hypothesis was then tested and confirmed by $^{15}N_2$ labeling experiments that were tracked by fluorescent *in situ* microscopy coupled to secondary ion mass spectrometry (FISH-SIMS) (Pernthaler et al. 2008). In cases where the process of interest involves cellular uptake of compounds, stable isotope probing is a powerful way to enrich functional guilds of interest prior to omics sequencing, thereby linking specific taxa and functions (Kalyuzhnaya et al. 2008; von Bergen et al. 2013).

References

Anantharaman, K., Breier, J. A. & Dick, G. J. (2016a) Metagenomic resolution of microbial functions in deep–sea hydrothermal plumes across the Eastern Lau Spreading Center. *ISME Journal*, **10**, 225–239.

Anantharaman, K., Brown, C. T., Hug, L. A., et al. (2016b) Thousands of microbial genomes shed light on interconnected biogeochemical processes in an aquifer system. *Nature Communications*, **7**, 13219.

Aristegui, J., Gasol, J. M., Duarte, C. M. & Herndl, G. J. (2009) Microbial oceanography of the dark ocean's pelagic realm. *Limnology and Oceanography*, **54**, 1501–1529.

Baric, R. S., Crosson, S., Damania, B., Miller, S. I. & Rubin, E. J. (2016) Next-generation high-throughput functional annotation of microbial genomes. *MBio*, **7**, e0145-16.

Battistuzzi, F. U., Feijao, A. & Hedges, S. B. (2004) A genomic timescale of prokaryote evolution: insights into the origin of methanogenesis, phototrophy, and the colonization of land. *BMC Evolutionary Biology*, **4**, 44.

Béjà, O., Aravind, L., Koonin, E. V., et al. (2000) Bacterial rhodopsin: evidence for a new type of phototrophy in the sea. *Science*, **289**, 1902–1906.

Béjà, O., Suzuki, M. T., Heidelberg, J. F., et al. (2002) Unsuspected diversity among marine aerobic anoxygenic phototrophs. *Nature*, **415**, 630–633.

Blank, C. E. & Sanchez-Baracaldo, P. (2010) Timing of morphological and ecological innovations in the cyanobacteria – a key to understanding the rise in atmospheric oxygen. *Geobiology*, **8**, 1–23.

Bolduc, B., Youens-Clark, K., Roux, S., Hurwitz, B. L. & Sullivan, M. B. (2017) iVirus: facilitating new insights in viral ecology with software and community data sets imbedded in a cyberinfrastructure. *ISME Journal*, **11**, 7–14.

Boyd, E. S., Anbar, A. D., Miller, S., Hamilton, T. L., Lavin, M. & Peters, J. W. (2011) A late methanogen origin for molybdenum-dependent nitrogenase. *Geobiology*, **9**, 221–232.

Boyd, E. S. & Peters, J. W. (2013) New insights into the evolutionary history of biological nitrogen fixation. *Frontiers in Microbiology*, **4**, 201.

Brocks, J. J. & Banfield, J. (2009) Unravelling ancient microbial history with community proteogenomics and lipid geochemistry. *Nature Reviews Microbiology*, **7**, 601–609.

Brown, C. T. & Tiedje, J. M. (2011) Metagenomics: the paths forward. In: F. de Bruijn (ed.), *Handbook of Molecular Microbial Ecology, Volume II: Metagenomics in Different Habitats*. John Wiley & Sons, Chichester.

Brown, C. T., Hug, L. A., Thomas, B. C., et al. (2015) Unusual biology across a group comprising more than 15% of domain Bacteria. *Nature*, **523**, 208–211.

Canfield, D. E., Stewart, F. J., Thamdrup, B., et al. (2010) A cryptic sulfur cycle in oxygen-minimum-zone waters off the Chilean coast. *Science*, **330**, 1375–1378.

Daims, H., Lebedeva, E. V., Pjevac, P., et al. (2015) Complete nitrification by Nitrospira bacteria. *Nature*, **528**, 504.

Daims, H., Lucker, S. & Wagner, M. (2016) A new perspective on microbes formerly known as nitrite-oxidizing bacteria. *Trends in Microbiology*, **24**, 699–712.

David, L. A. & Alm, E. J. (2011) Rapid evolutionary innovation during an Archaean genetic expansion. *Nature*, **469**, 93–96.

Deutschbauer, A., Price, M. N., Wetmore, K. M., et al. (2011) Evidence-based annotation of gene function in Shewanella oneidensis MR-1 using genome-wide fitness profiling across 121 conditions. *PloS Genetics*, **7**, e1002385.

Dick, G. J. & Lam, P. (2015) Omics approaches to microbial geochemistry. *Elements*, **11**, 403–408.

Dupont, C. L., Butcher, A., Valas, R. E., Bourne, P. E. & Caetano-Anolles, G. (2010) History of biological metal utilization inferred through phylogenomic analysis of protein structures. *Proceedings of the National Academy of Sciences of the United States of America*, **107**, 10567–10572.

Ettwig, K. F., Butler, M. K., Paslier, D. L., et al. (2010) Nitrite-driven anaerobic methane oxidation by oxygenic bacteria. *Nature*, **464**, 543–548.

Francis, C. A., Roberts, K. J., Beman, J. M., Santoro, A. E. & Oakley, B. B. (2005) Ubiquity and diversity of ammonia-oxidizing archaea in water columns and sediments of the ocean. *Proceedings of the National Academy of Sciences of the United States of America* **102**, 14683–14688.

Gilbert, J. A., Dick, G. J., Jenkins, B., et al. (2014) Meeting report: Ocean 'omics science, technology and cyberinfrastructure: current challenges and future requirements (August 20–23, 2013). *Standards in Genomic Sciences*, **9**, 1252–1258.

Glass, J. B. (2015) Microbes that meddle with metals. *Microbe*, **10**, 197–202.

Goodacre, N. F., Gerloff, D. L. & Uetz, P. (2014) Protein domains of unknown function are essential in bacteria. *mBio*, **5**, e00744–13.

Hallam, S. J., Putnam, N., Preston, C. M., et al. (2004) Reverse methanogenesis: testing the hypothesis with environmental genomics. *Science*, **305**, 1457–1462.

Handelsman, J. (2004) Metagenomics: application of genomics to uncultured microorganisms. *Microbiology and Molecular Biology Reviews*, **68**, 669–685.

Jansson, J. K. (2013) FORUM: Microbiology The life beneath our feet. *Nature*, **494**, 40–41.

Kalyuzhnaya, M. G., Lapidus, A., Ivanova, N., et al. (2008) High-resolution metagenomics targets specific functional types in complex microbial communities. *Nature Biotechnology*, **26**, 1029–1034.

Karl, D. M. (2002) Hidden in a sea of microbes. *Nature*, **415**, 590–591.

Könneke, M., Bernhard, A. E., de la Torre, J. R., Walker, C. B., Waterbury, J. B. & Stahl, D. A. (2005) Isolation of an autotrophic ammonia-oxidizing marine archaeon. *Nature*, **437**, 543–546.

Kraft, B., Tegetmeyer, H. E., Sharma, R., et al. (2014) The environmental controls that govern the end product of bacterial nitrate respiration. *Science*, **345**, 676–679.

Lam, P., Lavik, G., Jensen, M. M., et al. (2009) Revising the nitrogen cycle in the Peruvian oxygen minimum zone. *Proceedings of the National Academy of Sciences of the United States of America*, **106**, 4752–4757.

Mackelprang, R., Waldrop, M. P., Deangelis, K. M., et al. (2011) Metagenomic analysis of a permafrost microbial community reveals a rapid response to thaw. *Nature*, **480**, 368–371.

Markowitz, V. M., Mavromatis, K., Ivanova, N. N., Chen, I. M., Chu, K. & Kyrpides, N. C. (2009) IMG ER: a system for microbial genome annotation expert review and curation. *Bioinformatics*, **25**, 2271–2278.

Mason, O. U., Hazen, T. C., Borglin, S., et al. (2012) Metagenome, metatranscriptome and single-cell sequencing reveal microbial response to Deepwater Horizon oil spill. *ISME Journal*, **6**, 1715–1727.

Meyer, F., Paarmann, D., D'souza, M., et al. (2008) The metagenomics RAST server – a public resource for the automatic phylogenetic and functional analysis of metagenomes. *BMC Bioinformatics*, **9**, 386.

Moran, M. A. (2009) Metatranscriptomics: eavesdropping on complex microbial communities. *Microbe*, **4**, 329–335.

Newman, D. K., Orphan, V. J. & Reysenbach, A. L. (2012) Molecular biology's contributions to geobiology. In: A.H. Knoll & K.O. Konhauser (eds), *Fundamentals of Geobiology*. Blackwell Publishing, Chichester.

Oremland, R. S., Capone, D. G., Stolz, J. F. & Fuhrman, J. (2005) Whither or wither geomicrobiology in the era of 'community metagenomics'. *Nature*, **3**, 572–578.

Paez-Espino, D., Eloe-Fadrosh, E. A., Pavlopoulos, G. A., et al. (2016) Uncovering Earth's virome. *Nature*, **536**, 425–430.

Pearson, A. (2014) Lipidomics for geochemistry. *Treatise on Geochemistry, Second Edition*, **12**, 291–336.

Pernthaler, A., Dekas, A. E., Brown, C. T., Goffredi, S. K., Embaye, T. & Orphan, V. J. (2008) Diverse syntrophic partnerships from deep-sea methane vents revealed by direct cell capture and metagenomics. *Proceedings of the National Academy of Sciences of the United States of America*, **105**, 7052–7057.

Pinto, A. J., Marcus, D. N., Ijaz, U. Z., Bautista-de Lose Santos, Q. M., Dick, G. J. & Raskin, L. (2015) Metagenomic evidence for the presence of comammox Nitrospira-like bacteria in a drinking water system. *mSphere*, **1**, e00054-15.

Podar, M., Gilmour, C. C., Brandt, C. C., et al. (2015) Global prevalence and distribution of genes and microorganisms involved in mercury methylation. *Science Advances*, **1**, e1500675.

Pulquerio, M. J. F. & Nichols, R. A. (2007) Dates from the molecular clock: how wrong can we be? *Trends in Ecology & Evolution*, **22**, 180–184.

Ram, R. J., Verberkmoes, C., Thelen, M. P., et al. (2005) Community Proteomics of a Natural Microbial Biofilm. *Science*, **308**, 1915–1920.

Raymond, J. (2005) The evolution of biological carbon and nitrogen cycling – a genomic perspective. *Molecular Geomicrobiology*, **59**, 211–231.

Raymond, J. & Segre, D. (2006) The effect of oxygen on biochemical networks and the evolution of complex life. *Science*, **311**, 1764–1767.

Reed, D. C., Algar, C. K., Huber, J. A. & Dick, G. J. (2014) Gene-centric approach to integrating environmental genomics and biogeochemical models. *Proceedings of the National Academy of Sciences of the United States of America*, **111**, 1879–1884.

Reed, D. C., Breier, J. A., Jiang, H., et al. (2015) Predicting the response of the deep-ocean microbiome to geochemical perturbations by hydrothermal vents. *ISME Journal*, **9**, 1857–1869.

Rinke, C., Schwientek, P., Sczyrba, A., et al. (2013) Insights into the phylogeny and coding potential of microbial dark matter. *Nature*, **499**, 431–437.

Rivers, A. R., Sharma, S., Tringe, S. G., Martin, J., Joye, S. B. & Moran, M. A. (2013) Transcriptional response of bathypelagic marine bacterioplankton to the Deepwater Horizon oil spill. *ISME Journal*, **7**, 2315–2329.

Roberts, R. J., Karp, P., Kasif, S., Linn, S. & Buckley, M. R. (2004) *An Experimental Approach to Genome Annotation*. American Academy of Microbiology, Washington DC.

Rothman, D. H., Fournier, G. P., French, K. L., et al. (2014) Methanogenic burst in the end-Permian carbon cycle. *Proceedings of the National Academy of Sciences of the United States of America*, **111**, 5462–5467.

Saito, M. A., McIlvin, M. R., Moran, D. M., et al. (2014) Multiple nutrient stresses at intersecting Pacific Ocean biomes detected by protein biomarkers. *Science*, **345**, 1173–1177.

Saito, M. A., Sigman, D. M. & Morel, F. M. M. (2003) The bioinorganic chemistry of the ancient ocean: the co-evolution of cyanobacterial metal requirements and biogeochemical cycles at the Archean–Proterozoic boundary? *Inorganica Chimica Acta*, **356**, 308–318.

Schirrmeister, B. E., de Vos, J. M., Antonelli, A. & Bagheri, H. C. (2013) Evolution of multicellularity coincided with increased diversification of cyanobacteria and the Great Oxidation Event. *Proceedings of the National Academy of Sciences of the United States of America*, **110**, 1791–1796.

Seshadri, R., Kravitz, S. A., Smarr, L., Gilna, P. & Frazier, M. (2007) CAMERA: a community resource for metagenomics. *PLoS Biology*, **5**, e75.

Solden, L., Lloyd, K. G. & Wrighton, K. C. (2016) The bright side of microbial dark matter: lessons learned from the uncultivated majority. *Current Opinion in Microbiology*, **31**, 217–226.

Strous, M., Pelletier, E., Mangenot, S., et al. (2006) Deciphering the evolution and metabolism of an annamox bacterium from a community genome. *Nature*, **440**, 790–794.

Summons, R. E., Jahnke, L. L., Hope, J. M. & Logan, G. A. (1999) 2-Methylhopanoids as biomarkers for cyanobacterial oxygenic photosynthesis. *Nature*, **400**, 554–557.

Swan, B. K., Martinez-Garcia, M., Preston, C. M., et al. (2011) Potential for chemo-lithoautotrophy among ubiquitous bacteria lineages in the dark ocean. *Science*, **333**, 1296–12300.

Taupp, M., Mewis, K. & Hallam, S. J. (2011) The art and design of functional metagenomic screens. *Current Opinion in Biotechnology*, **22**, 465–472.

Temperton, B. & Giovannoni, S. J. (2012) Metagenomics: microbial diversity through a scratched lens. *Current Opinion in Microbiology*, **15**, 605–612.

Thompson, L. R., Zeng, Q., Kelly, L., et al. (2011) Phage auxiliary metabolic genes and the redirection of cyanobacterial host carbon metabolism. *Proceedings of the National Academy of Sciences of the United States of America*, **108**, E757–764.

Trembath-Reichert, E., Wilson, J. P., Mcglynn, S. E. & Fischer, W. W. (2015) Four hundred million years of silica biomineralization in land plants. *Proceedings of the National Academy of Sciences of the United States of America*, **112**, 5449–5454.

Tsementzi, D., Wu, J., Deutsch, S., et al. (2016) SAR11 bacteria linked to ocean anoxia and nitrogen loss. *Nature*, **536**, 179–183.

Tyson, G. W., Lo, I., Baker, B. J., Allen, E. E., Hugenholtz, P. & Banfield, J. F. (2005) Genome-directed isolation of the key nitrogen fixer *Leptospirillum ferrodiazotrophum* sp. nov. from an acidophilic microbial community. *Applied and Environmental Microbiology*, **71**, 6319–6324.

Uchiyama, T., Abe, T., Ikemura, T. & Watanabe, K. (2005) Substrate-induced gene-expression screening of environmental metagenome libraries for isolation of catabolic genes. *Nature Biotechnology*, **23**, 88–93.

Van Kessel, M., Speth, D. R., Albertsen, M., et al. (2015) Complete nitrification by a single microorganism. *Nature*, **528**, 555–559.

Venter, J. C., Remington, K., Heidelberg, J. F., et al. (2004) Environmental genome shotgun sequencing of the Sargasso Sea. *Science*, **304**, 66–74.

Von Bergen, M., Jehmlich, N., Taubert, M., et al. (2013) Insights from quantitative metaproteomics and protein-stable isotope probing into microbial ecology. *ISME Journal*, **7**, 1877–1885.

Walsh, D. A., Zaikova, E., Howes, C. G., et al. (2009) Metagenome of a versatile chemolithoautotroph from expanding oceanic dead zones. *Science*, **326**, 578–582.

Ward, B. B., Devol, A. H., Rich, J. J., et al. (2009) Denitrification as the dominant nitrogen loss process in the Arabian Sea. *Nature*, **461**, 78–81.

Welander, P. V., Coleman, M. L., Sessions, A. L., Summons, R. E. & Newman, D. K. (2010) Identification of a methylase required for 2-methylhopanoid production and implications for the interpretation of sedimentary hopanes. *Proceedings of the National Academy of Sciences of the United States of America*, **107**, 8537–8542.

Wilbanks, E. G., Jaekel, U., Salman, V., et al. (2014) Microscale sulfur cycling in the phototrophic pink berry consortia of the Sippewissett Salt Marsh. *Environmental Microbiology*, **16**, 3398–3415.

Wilkins, M. J., Verberkmoes, N. C., Williams, K. H., et al. (2009) Proteogenomic monitoring of *Geobacter* physiology during stimulated uranium bioremediation. *Applied and Environmental Microbiology*, **75**, 6591–6599.

Wrighton, K. C., Thomas, B. C., Miller, C. S., et al. (2012) Fermentation, hydrogen, and sulfur metabolism in multiple uncultivated bacterial phyla. *Science*, **337**, 1661–1665.

Wrighton, K. C., Castelle, C. J., Wilkins, M. J., et al. (2014) Metabolic interdependencies between phylogenetically novel fermenters and respiratory organisms in an unconfined aquifer. *ISME Journal*, **8**, 1452–1463.

Wrighton, K. C., Castelle, C. J., Varaljay, V. A., et al. (2016) RubisCO of a nucleotide pathway from Archaea is found in diverse uncultivated phyla in bacteria. *ISME Journal*, **10**, 2702–2714.

Yooseph, S., Sutton, G., Rusch, D. B., et al. (2007) The *Sorcerer II* Global Ocean Sampling expedition: expanding the universe of protein families. *PLoS Biology*, **5**, e16.

Zerkle, A. L., House, C. H. & Brantley, S. L. (2005) Biogeochemical signatures through time as inferred from whole microbial genomes. *American Journal of Science*, **305**, 467–502.

Overview of Approaches

From Whole-Community Shotgun Sequencing to Single-Cell Genomics

Introduction

Omics approaches can now be applied in a variety of ways to study microorganisms in the environment. The materials that can be used as a starting point for extraction of nucleic acids include whole microbial communities sampled directly from the environment, enrichments or pure cultures, and populations or even single cells that have been separated from other members of their communities by physical means (Fig. 4.1). Here our discussion focuses on genomics. In theory, each of these methods can also be applied to transcriptomics and proteomics, but in practice some of them (e.g., single-cell genomics) are currently not applicable to transcriptomics and proteomics due to limited material.

4.1 Choosing the Right Approach

4.1.1 Whole-Community Approaches

Because DNA sequencing technologies now offer massive throughput at low cost, whole microbial communities can be reasonably characterized by shotgun sequencing. DNA, RNA, and/or protein is extracted directly from whole environmental samples (e.g., cells collected from water by filtration; whole soil samples) and randomly fragmented and sequenced (see Fig. 4.1). This approach can be appealing for those who want a relatively unbiased view of the genetic and functional potential and expression of whole communities. For example, such views can offer insights into how different populations, functions, and trophic levels (i.e., grazers, autotrophs, heterotrophs,

Genomic Approaches in Earth and Environmental Sciences, First Edition. Gregory Dick.

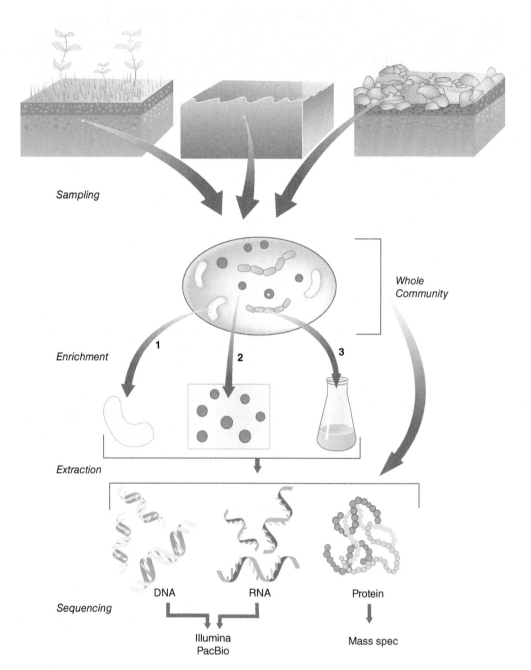

Figure 4.1 Overview of approaches and procedures for omics approaches to the Earth and environmental sciences. Sampling is conducted from various aquatic and terrestrial environments. Omics studies can be performed on the whole microbial community (*far right*) or on specific portions of the community that are targeted, for example, by (1) single-cell approaches, (2) enrichment of populations by techniques such as flow cytometry, or (3) isolation of pure or mixed cultures.

viruses) vary temporally, spatially, or as a function of environmental conditions. A notable strength of this approach is the potential for discovery; the history of microbiology shows that many of the greatest advances have been serendipitous (Jansson & Prosser 2013); our knowledge of the microbial world is so poor that often we do not know the right questions to ask.

The challenge of whole-community shotgun omics is the size and complexity of the datasets produced, especially for diverse microbial communities such as those found in soil. Key issues in considering the whole-community approach and how best to analyze the data are the availability of reference genomes and the feasibility of *de novo* assembly (see section 4.2.2 and Chapter 6). While *de novo* assembly was once possible only for simple communities, recent advances enable its application to highly complex communities (Anantharaman et al. 2016; Sharon & Banfield 2013). In addition, the cost and time saved by the simple presequencing steps of shotgun sequencing relative to more complicated procedures for approaches such as single-cell genomics should be considered.

4.1.2   Targeted Approaches: Physical, Microbiological, and Isotopic Enrichment

When specific populations or functions are of interest, sequencing the whole community may not be necessary. Moreover, targeted capture of specific microbial groups followed by omics can be a powerful way to identify the microbes and/or genes, enzymes, and pathways that underpin processes of interest. This overall notion is as old as microbiology itself; microorganisms are isolated or enriched on the basis of metabolic or physiological characteristics. Enrichment and isolation of specific microbial species or consortia prior to omics analysis can greatly reduce the complexity of the dataset and the required sequencing effort (Delmont et al. 2015) (see Fig. 4.1). In cases where pure cultures can be obtained, longer sequencing reads offered by technologies such as Illumina MiSeq offer a cost-effective way to obtain high-quality and even complete genomes (Coil et al. 2015), whereas extremely long reads can resolve repeat regions in complex genomes that are otherwise difficult to assemble (Chin et al. 2013).

For microbial groups that cannot be isolated or sufficiently enriched, stable isotope probing can be used to enrich for macromolecules that incorporate elements through assimilatory pathways of interest. For example, to identify methanotrophic populations, ^{13}C-labeled methane can be fed to a community (Kalyuzhnaya et al. 2008). Extracted DNA or RNA that has incorporated the labeled carbon can then be fractionated on the basis of density such that the heavy nucleic acids containing ^{13}C are physically separated from the rest of the community members that did not take up the compound of interest, and the heavy fraction can be sequenced to identify the associated microbes and genes.

Another method to target specific populations for omics approaches is *physical selection* (see Fig. 4.1). This is commonly done by oceanographers via flow cytometry, in which cells of a specific size class and/or natural fluorescence (due to pigment content) are identified and sorted, resulting in a highly enriched population for omics analysis (Zehr et al. 2008). Alternatively, genetic probes can be used to "label" specific taxa via fluorescence-activated cell sorting (Martinez-Garcia et al. 2012; Pernthaler et al. 2008; Woyke et al. 2009). Such approaches greatly reduce the complexity of the assemblage to be sequenced while still providing insights into population-level diversity afforded by community sequencing.

When a certain size class of cell is of interest, the community can be fractionated by analyzing the appropriate size fraction after water is passed through a series of filtration steps (Satinsky et al. 2014; Venter et al. 2004). For example, in aquatic systems, one might focus on viruses (<0.2 μm), free-living bacteria (>0.2 μm, <3 μm), or particle attached bacteria (>3 μm). Elegant downstream methods for efficiently recovering the viral fraction have now been developed (Duhaime et al. 2012). In studies of symbionts where material from host cells is an issue, it may be necessary to physically separate symbiotic microbial cells from those of the host, e.g., via Percoll gradients (Caro et al. 2007). While size fractionation can be effective at separating major groups, reducing the complexity of datasets, and providing efficient access to the target of interest, such methods can complicate analytical efforts by increasing the number of datasets to be analyzed (where multiple fractions are studied).

4.1.3 Single-Cell Genomics

The finest-scale form of physical selection is *single-cell genomics*, in which individual microbial cells are isolated and the genome is sequenced. This can be accomplished either by microfluidics and laser tweezers or by fluorescence activated cell sorting (Blainey and Quake, 2014; Gawad et al. 2016; Stepanauskas, 2012; Xu et al. 2016). Single-cell genomics provides unparalleled views of the genomic content at the level of individual cells (Gawad et al. 2016; Hedlund et al. 2014; Stepanauskas 2012; Woyke et al. 2009). Single-cell genomics can effectively and definitively connect phylogeny and function (Rinke et al. 2013; Stepanauskas 2012; Swan et al. 2011), whereas whole-community shotgun methods may not yield sufficient assemblies to establish this link (though they often do). An important advantage of this approach is that the risk of chimeric assemblies is minimized (Stepanauskas 2012), though chimeras are still possible due to repeats *within* genomes. Single-cell genomics can reveal physical associations such as those between viruses (Roux et al. 2014) and endosymbionts (Yoon et al. 2011) and their hosts. These relationships may not be apparent through whole-community sequencing. Another advantage of single-cell genomics is that it preserves information on how different gene variants are linked within a genome; this information is often lost via metagenomics if strains cannot be resolved by advanced binning procedures (see section 6.5).

A persistent challenge of single-cell genomics is that the genome amplification necessary to generate enough material for sequencing leads to highly uneven coverage of the genome, which often prevents sequencing and assembly of complete genomes (Sharon & Banfield 2013). Genomes derived from single cells and cultures are valuable for interpretation of whole-community omics datasets, and the whole-community data can provide quantitative information, so these approaches are complementary (Hedlund et al. 2014).

4.2 Experimental Design and Sampling Considerations

Perhaps the most important phase of any microbial community omics study occurs *before* any samples are collected; careful consideration of the experimental design is critical. Several recent reviews have discussed various aspects of this issue in depth (Muller et al. 2013; Thomas et al. 2012, 2015; Nayfach & Pollard 2016). Paramount are decisions of how many samples to sequence and how much sequencing effort to devote to each sample. Below, we touch on some of the major issues that inform this decision, including replication, sequencing effort, and costs. It is also necessary to consider the very real concerns of extraction bias and contamination. The previous section on size fractionation and a later section on statistical analyses are also relevant here. Another important aspect of experimental design for environmental omics is the establishment of procedures for collecting and archiving metadata and samples in forms that are easily disseminated.

4.2.1 Replication

Too often, omics studies do not analyze a sufficient number of replicates in order to allow statistically rigorous analyses (Knight et al. 2012; Nayfach & Pollard 2016; Prosser 2010). While replicates may not be necessary for qualitative studies (e.g., reconstructing genomes to uncover the metabolic potential of a population), replication is required for most quantitative studies (e.g., comparing gene abundance or expression across samples). Indeed, it has been shown that variation due to technical variation may in some cases be comparable to the variation due to biological variation (McCarthy et al. 2015; Tsementzi et al. 2014). Hence, technical replicates are important to assess uncertainty due to technical variation (e.g., between different DNA extractions or sequencing runs, even when using the same technology). It is also important to have "biological replicates" (i.e., different samples or subsamples of the same experimental treatment or field condition) so that the robustness of conclusions can be assessed. Fortunately, the dramatic decline in sequencing costs makes replication feasible.

4.2.2 Estimating Sequencing Effort: How Much Sequencing Do I Need to Do?

How much sequencing effort is required? The answer to this question depends on several factors.

First, what is the goal of the sequencing effort? To recover complete genomes from dominant community members? To recover complete genomes from a particular functional group of organisms that may not be abundant? To profile the major functions of the community? Sequencing effort must be tuned to the objectives at hand. Second, what is the biological diversity of the sample? More diverse samples demand greater sequencing effort to cover all those different genotypes. Instead of trying to estimate the total number of genotypes, a crude yet reasonable approach is to base the calculation on the organism of interest that is least abundant (so that all other target organisms will be "sampled" at least as well as the least abundant one). The key here is to estimate the fraction of total community DNA represented by the least abundant organism; this can be difficult but a crude starting point is to assume that it is equal to the relative abundance in 16S datasets (in reality, this fraction will be further influenced by 16S gene copy number and genome size relative to genome sizes of other community members). Hence, a simple calculation to estimate sequencing effort is:

$$\frac{([\text{genome size}] * [\text{desired coverage}])}{(\text{relative abundance of least abundant target organism as fraction of community DNA})}$$

Let's consider a simple example. We have a community in which we have identified 100 operational taxonomic units (OTUs) that each represent at least 0.1% of the total community (the threshold of abundance that we are interested in here). Let's assume that we want to achieve 100× genomic coverage of all OTUs present at this abundance level, that each of these OTUs contains just one genotype (probably not valid, see below), and that the average genome size of community members is relatively consistent at about 5 Mb. In order to achieve 100 × coverage of those OTUs that represent 0.1% community abundance, we need to sequence 5 Mb * 100 (target coverage) * 1000 (0.1% = 0.001/1 = 1000) = 500 Gbp.

In reality, it is much more complicated to accurately determine the amount of sequencing required to achieve desired outcomes. What really matters for genome assembly is the genomic diversity of populations and communities, especially the frequency of repeat sequences, and variation of gene order and orientation between genotypes. Because such genomic diversity can occur at very fine levels of divergence, 16S rRNA data often provides only limited constraints. For example, in one of our studies some OTUs as defined at the 97% 16S rRNA gene identify level assembled neatly into one or several genomes, whereas the assemblies of other OTUs (that were even more abundant at the 16S level!) were fragmented into short contigs (Lesniewski et al. 2012). Such fragmentation can be due to either intragenome complexity or intra-OTU genomic diversity.

There have been advances beyond such intuitive yet crude "back of the envelope" calculations. Wendl et al. (2013) developed formalized methods by extending the Lander–Waterman theory (Lander & Waterman 1988) to statistically model metagenomic coverage as a function of sequencing effort, but these methods have not yet been widely applied to real metagenomic sequencing projects. Where genome assembly is not the goal, metrics and methods have been developed to estimate the coverage of a metagenomic dataset and the amount of sequencing needed to cover the total diversity in a sample (Rodriguez & Konstantinidis 2014). Note that "coverage" here refers to the fraction of diversity sampled rather than contig read-depth (the latter is the main way it is used in this book). This has been suggested as a useful metric to report for metagenomic studies (Rodriguez & Konstantinidis 2014).

Three additional practical issues should be kept in mind when determining how much sequencing to conduct. First, beyond a certain threshold increased sequencing depth can actually inhibit genomic assembly due to effects of sequencing errors (Lonardi et al. 2015). Second, more broadly, we should recognize that bioinformatics analysis is now more expensive than sequencing; because this new paradigm has been slow to sink in, projects often do not have sufficient bioinformatics resources relative to sequence data (Sboner et al. 2011). Finally, given the still substantial costs of omics analyses (including bioinformatics) and the frequent need to analyze numerous samples for comparison and/or replication, consider a tiered approach in which cheaper/higher-throughput methods are used first to screen a large number of samples, then that data is used to pick key samples for further omics analyses (Tickle et al. 2013).

4.2.3   From Sample to Data: Biases Due to Preservation, Storage, Extraction, and Sequencing

Although culture-independent molecular approaches to microbial ecology avoid the well-known biases of cultivation, they are subject to a variety of other biases associated with how samples are preserved, stored, extracted, and sequenced (Forney et al. 2004; McCarthy et al. 2015; Nayfach & Pollard 2016; Temperton & Giovannoni 2012; Tsementzi et al. 2014).

A particularly important source of potential bias is the method used for extraction of nucleic acids. Different extraction methods vary in their effectiveness of cell lysis and DNA recovery. This effectiveness can even vary between different microbial groups, presumably due to differences in the properties of cell walls and membranes. While these technical effects are typically small relative to the large biological differences often observed between samples or treatments, they may be substantial where subtle differences or specific microbial groups are of interest (McCarthy et al. 2015). The optimal method depends on the nature of the sample, the question being asked, and the downstream applications. If difficult-to-lyse cells are the target, or are abundant in the community, a physical method such as bead beating may be required. However, our lab has found that although DNA obtained with a bead-beating method is fine for Illumina sequencing, it does not work for PacBio sequencing, presumably due to damage to the DNA (e.g., nicks and shearing). In addition, special extraction or separation methods may be needed to avoid inhibitors or enzymes in certain sample types such as soil (Delmont et al. 2011). When multiple omics methods (e.g., genomics, transcriptomics, proteomics) are being conducted and compared (e.g., gene expression as a ratio of RNA:DNA copies), it is important that the same extraction procedures are used where possible (McCarthy et al. 2015; Muller et al. 2013). This can be achieved by using a single-stream extraction like the Qiagen AllPrep kit, or by customizing other extraction methods to use a general lysis procedure prior to DNA and RNA isolation procedures.

Environmental and geobiological studies often involve analysis of samples with low biomass, and these samples may be in limited supply due to accessibility issues (e.g., the deep sea or terrestrial subsurface). Such low-biomass samples present several challenges. First, many DNA sequencing technologies require substantial inputs of material for construction of sequencing libraries (hundreds of nanograms to micrograms). Until recently, lower DNA concentrations would have to be increased with whole-genome amplification methods such as multiple displacement amplification, which have substantial biases due to uneven amplification (Abbai et al. 2012; Pinard et al. 2006). Newer methods of sequencing library construction amplify and multiplex DNA via polymerase chain reaction, which also introduces biases (Aird et al. 2011; Kozarewa et al. 2009; van Nieuwerburgh et al. 2011). Commercial kits for preparation of metatranscriptomic sequencing libraries such as ScriptSeq v2 (Illumina) need as little as 500 pg of starting material. However, there are apparent biases across the different methods for library preparation; as of 2015, the currently available methods had not been rigorously evaluated and compared, but it is clear that metatranscriptomes prepared using different methods may not be comparable (Alberti et al. 2014). Additional issues specific to metatranscriptomics, such as sample preservation and rRNA removal, are discussed further in Chapter 9.

Another challenge that is exacerbated in low-biomass samples is contamination. Recent studies show that contamination is a pervasive problem in DNA sequence datasets from microbial communities (Breier et al. 2014; Salter et al. 2014; Tanner et al. 1998). This contamination can come from many sources, including DNA extraction kits, laboratory reagents, instruments and containers used for sampling, and even the researcher (Salter et al. 2014; Weiss et al. 2014). Not surprisingly, low-biomass samples are more susceptible to contamination due to a reduced signal to noise ratio; when there is less real sample ("signal"), the fraction of the total sample contributed by contaminants ("noise") is greater. These studies highlight the important of negative controls, replicates, and benchmarking and validation of protocols.

4.2.4   Estimating Absolute Abundance with Internal Standards

A key issue with omics data is that it typically comes in the form of relative abundance. Interpreting comparisons of relative abundance data across samples is complicated by the fact that the results of any entity (gene or organism) are affected by abundance of other entities in that sample. For example, consider a study comparing a "normal" state (sample A) to a "bloom" state in which certain organisms have rapidly proliferated (sample B). An organism whose actual abundance remains the same between both states would have a lower *relative abundance* in sample B due to the increased abundance of other community members. Hence, it is often desirable to produce data in terms of *absolute abundance*. Efficient methods for calculating absolute abundance from shotgun omics data have been developed (Moran et al. 2013; Satinsky et al. 2013) but are not yet as widely applied as they should be.

4.3   Overview of Current DNA Sequencing Technologies

DNA sequencing technologies have enabled the omics revolution and continue to shape the way in which omics projects are approached today. A number of options for DNA sequencing are currently available, each offering advantages and disadvantages with regard to cost, throughput, sequence read length, quality, and biases (Table 4.1). For a historical perspective, here we also include technologies that are no longer widely used. Several recent reviews provide more in-depth discussion of these technologies (Escobar-Zepeda et al. 2015; Koren & Phillippy 2015; Loman & Watson 2015; Loman et al. 2012; Nagarajan & Pop 2013; Thomas et al. 2012; Weinstock 2012).

Sanger sequencing was the workhorse of early microbial genomics and it still provides high-quality, relatively long sequence reads that remain useful for some purposes. However, due to its much higher relative cost, it is no longer used for high-throughput applications involving whole genomes. 454 Pyrosequencing was one of the first so-called "next-generation" technologies, and it had a substantial impact on microbial

Table 4.1 DNA sequencing technologies.*

Technology	Error rate (%)	Read length (bp)	Reads per run	Cost per Gb ($USD)
Sanger	<0.1	800	96	2 000 000
454	1	400–1000	1 000 000	10 000
Illumina HiSeq (per lane)	0.1	100–150	300 000 000	30
Illumina MiSeq	0.1	150–300	40 000 000	150
Illumina TruSeq Synthetic Long-Reads	0.1	8000–10 000	Variable	Variable
Oxford Nanopore	3–5	5000–15 000+ (up to 100s kb)	Variable	Variable
PacBio	13[†]	1000–40 000	50 000	500
IonTorrent	1	400	4 000 000	500

*Data provided are approximate and are changing with rapidly advancing technologies.
†For a single read, not consensus; see text.
Data taken from Nagajaran and Pop (2013), Escobar-Zepeda et al. (2015) and Genohub (2017).

omics in the mid 2000s. Two systematic issues of 454 were homopolymers (Huse et al. 2007), which cause errors in gene calling due to frameshifts, and artificial replicates, which cause overestimates of gene and taxon abundance (Gomez-Alvarez et al. 2009). 454 was superseded by the much cheaper and higher throughput capabilities of Illumina technologies. The Illumina HiSeq (2000/2500/3000/4000) provides unparalleled throughput and low cost, while the MiSeq provides longer read lengths in a platform that is more manageable and affordable for smaller labs. Like 454, all Illumina technologies also typically require PCR amplification with universal primers and subsequent cluster amplification, which introduces potential biases, especially at extremes of nucleotide composition (Aird et al. 2011; Kozarewa et al. 2009; Minoche et al. 2011; Schirmer et al. 2015). Although some of these issues have been mitigated by optimized protocols, they still exist.

PacBio sequencing offers exceptionally long read lengths (10–40 kb), and methods for dealing with its high error rates (~15%) have now been developed (Berlin et al. 2015; Koren et al. 2012; Liao et al. 2015). Although PacBio is not competitive with Illumina in terms of cost, it is incredibly valuable for assembling large microbial genomes with high complexity. Although not yet feasible for standalone omics analysis of complex microbial communities, PacBio is valuable for less complex communities or when employed in tandem with higher throughput, higher quality technologies such as Illumina (Frank et al. 2016). The homolog blog provides a nice overview of PacBio sequencing (www.homolog.us/Tutorials/index.php?p=1.1&s=5).

Another approach, taken by Illumina TruSeq synthetic long-read sequencing, is to use a modified library preparation in which regular short reads are reassembled into synthetic long reads (Kuleshov et al. 2014). Its application to metagenomics has proven useful for resolving intraspecies diversity (Kuleshov et al. 2016; Sharon et al. 2015). In some cases, the best approach may in fact be to combine two different technologies with complementary strengths, that is, a hybrid of Illumina and PacBio. An additional consideration is that short-read technologies offer a variety of insert sizes for paired end sequencing, which can assist in metagenomic assembly. Insert sizes are sufficiently short that the reads overlap, allowing paired reads to be merged, yielding longer effective read lengths (Liu et al. 2012; Seemann 2012). Additional technologies such as Ion Torrent have been developed (Loman & Watson 2015) but have yet to be demonstrated as competitive for the purpose of metagenomics.

4.4 Quality Control and Sequence Processing

Ensuring the quality of DNA sequence is critical for maintaining the integrity of many downstream analyses. Next-generation sequencing platforms, while producing massive data in a cost-effective manner, have a number of known biases and limitations that must be taken into account. The steps of sequence quality control, detailed in sections below, typically include (i) using the information held within a FASTQ file, (ii) dereplicating the data to save computational cost and/or remove artifacts, (iii) trimming bad sequence from reads and completely removing reads below some threshold prior to downstream approaches such as assembly (Fig. 4.2). FASTQ files are the most commonly used format for describing sequence quality. The format and an example of FASTQ files are shown in Figure 4.3.

4.4.1 Dereplication

A common practice in the processing of shotgun sequencing datasets is to dereplicate the data, i.e., remove reads that share identical sequence and start and stop positions. The rationale is two-fold:

- avoiding computational costs of downstream analyses by removing reads that provide no new information
- the assumption that duplicated reads are artifacts of the sequencing library construction process (Gomez-Alvarez et al. 2009).

Figure 4.2 Pipeline for quality control of next-generation sequence data.

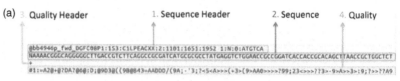

Figure 4.3 (a) The FASTQ format; (b) screenshot of a FASTQ file.

The latter stems from the notion that when DNA or cDNA is randomly fragmented, the likelihood of fragmentation at exactly the same genomic positions should be exceedingly low. Such duplicate reads can represent a significant portion of Illumina datasets (Aird et al. 2011) and may be produced as artifacts during the library construction process (Kozarewa et al. 2009). However, at high-sequencing depth, duplicated reads are possible due to coincidence rather than artifact, and dereplication of the data may overcorrect amplification bias, leading to artifacts in itself (Zhou et al. 2014). The possibility that duplicate reads are real is higher for short transcripts with defined start and stop sites. Although models for correcting for such artifacts while considering the possibility that some duplicates could be real have been developed (Zhou et al. 2014), they are not yet widely applied to microbial community omics data.

4.4.2 Trimming

Trimming is used to remove positions of DNA sequence that are of questionable quality. Such low-quality sequence can interfere with numerous downstream tasks such as assembly, mapping reads to reference genomes, or querying databases. For Illumina sequencing, a quality score of 20 or above is considered acceptable. As an example, a simple trimming algorithm might look something like this.

```
Take window of N bases
Average quality score, A, for this window
Check if A < Desired minimum quality
        if yes → chop sequence here
                Go to next sequence
        if no → Step by S bases
Repeat
```

Sickle is an effective and commonly used method for trimming and it is freely available (https://github.com/najoshi/sickle). Methods of error correction and merging overlapping reads also reduce error rates (Schirmer et al. 2015).

References

Abbai, N. S., Govender, A., Shaik, R. & Pillay, B. (2012) Pyrosequence analysis of unamplified and whole genome amplified DNA from hydrocarbon-contaminated groundwater. *Molecular Biotechnology*, **50**, 39–48.

Aird, D., Ross, M. G., Chen, W. S., et al. (2011) Analyzing and minimizing PCR amplification bias in Illumina sequencing libraries. *Genome Biology*, **12**, R18.

Alberti, A., Belser, C., Engelen, S., et al. (2014) Comparison of library preparation methods reveals their impact on interpretation of metatranscriptomic data. *BMC Genomics*, **15**, 912.

Anantharaman, K., Brown, C. T., Hug, L. A., et al. (2016) Thousands of microbial genomes shed light on interconnected biogeochemical processes in an aquifer system. *Nature Communications*, **7**, 13219.

Berlin, K., Koren, S., Chin, C. S., Drake, J. P., Landolin, J. M. & Phillippy, A. M. (2015) Assembling large genomes with single-molecule sequencing and locality-sensitive hashing. *Nature Biotechnology*, **33**, 623–630.

Blainey, P. C. & Quake, S. R. (2014) Dissecting genomic diversity, one cell at a time. *Nature Methods*, **11**, 19–21.

Breier, J. A., Gomez-Ibanez, D. A., Sayre-McCord, R. T., et al. (2014) A large volume particulate and water multi-sampler with in situ preservation for microbial and biogeochemical studies. *Deep-Sea Research I*, **94**, 195–206.

Caro, A., Gros, O., Got, P., de Wit, R. & Trousseltier, M. (2007) Characterization of the population of the sulfur-oxidizing symbiont of Codakia orbicularis (Bivalvia, Lucinidae) by single-cell analyses. *Applied and Environmental Microbiology*, **73**, 2101–2109.

Chin, C. S., Alexander, D. H., Marks, P., et al. (2013) Nonhybrid, finished microbial genome assemblies from long-read SMRT sequencing data. *Nature Methods*, **10**, 563–569.

Coil, D., Jospin, G. & Darling, A. E. (2015) A5-miseq: an updated pipeline to assemble microbial genomes from Illumina MiSeq data. *Bioinformatics*, **31**, 587–589.

Delmont, T. O., Robe, P., Clark, I., Simonet, P. & Vogel, T. M. (2011) Metagenomic comparison of direct and indirect soil DNA extraction approaches. *Journal of Microbiological Methods*, **86**, 397–400.

Delmont, T. O., Eren, A. M., Maccario, L., et al. (2015) Reconstructing rare soil microbial genomes using in situ enrichments and metagenomics. *Frontiers in Microbiology*, **6**, 358.

Duhaime, M. B., Deng, L., Poulos, B. T. & Sullivan, M. B. (2012) Towards quantitative metagenomics of wild viruses and other ultra-low concentration DNA samples: a rigorous assessment and optimization of the linker amplification method. *Environmental Microbiology*, **14**, 2526–2537.

Escobar-Zepeda, A., de Leon, A. V. P. & Sanchez-Flores, A. (2015) The road to metagenomics: from microbiology to DNA sequencing technologies and bioinformatics. *Frontiers in Genetics*, **6**, 348.

Forney, L. J., Zhou, X. & Brown, C. J. (2004) Molecular microbial ecology: land of the one-eyed king. *Current Opinion in Microbiology*, **7**, 210–220.

Frank, J. A., Pan, Y., Tooming-Klunderud, A., et al. (2016) Improved metagenome assemblies and taxonomic binning using long-read circular consensus sequence data. *Science Reports*, **6**, 25373.

Gawad, C., Koh, W. & Quake, S. R. (2016) Single-cell genome sequencing: current state of the science. *Nature Reviews Genetics*, **17**, 175–188.

Genohub (2017) Choosing the Right NGS Sequencing Instrument for Your Study. Available at: https://genohub.com/ngs-instrument-guide/ (accessed 30 October 2017).

Gomez-Alvarez, V., Teal, T. K. & Schmidt, T. M. (2009) Systematic artifacts in metagenomes from complex microbial communities. *ISME Journal*, **3**, 1314–1317.

Hedlund, B. P., Dodsworth, J. A., Murugapiran, S. K., Rinke, C. & Woyke, T. (2014) Impact of single-cell genomics and metagenomics on the emerging view of extremophile "microbial dark matter". *Extremophiles*, **18**, 865–875.

Huse, S. M., Huber, J. A., Morrison, H. G., Sogin, M. L. & Welch, D. M. (2007) Accuracy and quality of massively parallel DNA pyrosequencing. *Genome Biology*, **8**, R143.

Jansson, J. & Prosser, J. I. (2013) The life beneath our feet. *Nature*, **494**, 40–41.

Kalyuzhnaya, M. G., Lapidus, A., Ivanova, N., et al. (2008) High-resolution metagenomics targets specific functional types in complex microbial communities. *Nature Biotechnology*, **26**, 1029–1034.

Knight, R., Jansson, J., Field, D., et al. (2012) Unlocking the potential of metagenomics through replicated experimental design. *Nature Biotechnology*, **30**, 513–520.

Koren, S. & Phillippy, A. M. (2015) One chromosome, one contig: complete microbial genomes from long-read sequencing and assembly. *Current Opinion in Microbiology*, **23**, 110–120.

Koren, S., Schatz, M. C., Walenz, B. P., et al. (2012) Hybrid error correction and de novo assembly of single-molecule sequencing reads. *Nature Biotechnology*, **30**, 693–700.

Kozarewa, I., Ning, Z., Quail, M. A., Sanders, M. J., Berriman, M. & Turner, D. J. (2009) Amplification-free Illumina sequencing-library preparation facilitates improved mapping and assembly of (G+C)-biased genomes. *Nature Methods*, **6**, 291–295.

Kuleshov, V., Xie, D., Chen, R., et al. (2014) Whole-genome haplotyping using long reads and statistical methods. *Nature Biotechnology*, **32**, 261–266.

Kuleshov, V., Jiang, C., Zhou, W., Jahanbani, F., Batzoglou, S. & Snyder, M. (2016) Synthetic long-read sequencing reveals intraspecies diversity in the human microbiome. *Nature Biotechnology*, **34**, 64–69.

Lander, E. S. & Waterman, M. S. (1988) Genomic mapping by fingerprinting random clones: a mathematical analysis. *Genomics*, **2**, 231–239.

Lesniewski, R. A., Jain, S., Anantharaman, K., Schloss, P. D. & Dick, G. J. (2012) The metatranscriptome of a deep-sea hydrothermal plume is dominated by water column methanotrophs and lithotrophs. *ISME Journal*, **6**, 2257–2268.

Liao, Y. C., Lin, S. H. & Lin, H. H. (2015) Completing bacterial genome assemblies: strategy and performance comparisons. *Science Reports*, **5**, 8747.

Liu, B., Yuan, J., Yiu, S. M., et al. (2012) COPE: an accurate k-mer-based pair-end reads connection tool to facilitate genome assembly. *Bioinformatics*, **28**, 2870–2874.

Loman, N. J. & Watson, M. (2015) Successful test launch for nanopore sequencing. *Nature Methods*, **12**, 303–304.

Loman, N. J., Constantinidou, C., Chan, J. Z., et al. (2012) High-throughput bacterial genome sequencing: an embarrassment of choice, a world of opportunity. *Nature Reviews Microbiology*, **10**, 599–606.

Lonardi, S., Mirebrahim, H., Wanamaker, S., et al. (2015) When less is more: "slicing" sequencing data improves read decoding accuracy and de novo assembly quality. *Bioinformatics*, **31**, 2972–2980

Martinez-Garcia, M., Brazel, D. M., Swan, B. K., et al. (2012) Capturing single cell genomes of active polysaccharide degraders: an unexpected contribution of Verrucomicrobia. *PLoS One*, **7**, e35314.

McCarthy, A., Chiang, E., Schmidt, M. L. & Denef, V. J. (2015) RNA preservation agents and nucleic acid extraction method bias perceived bacterial community composition. *PLoS One*, **10**, e0121659.

Minoche, A. E., Dohm, J. C. & Himmelbauer, H. (2011) Evaluation of genomic high-throughput sequencing data generated on Illumina HiSeq and genome analyzer systems. *Genome Biology*, **12**, R112.

Moran, M. A., Satinsky, B., Gifford, S. M., et al. (2013) Sizing up metatranscriptomics. *ISME Journal*, **7**, 237–243.

Muller, E. E., Glaab, E., May, P., Vlassis, N. & Wilmes, P. (2013) Condensing the omics fog of microbial communities. *Trends in Microbiology*, **21**, 325–333.

Nagarajan, N. & Pop, M. (2013) Sequence assembly demystified. *Nature Reviews Genetics*, **14**, 157–167.

Nayfach, S. & Pollard, K. S. (2016) Toward accurate and quantitative comparative metagenomics. *Cell*, **166**, 1103–1116.

Pernthaler, A., Dekas, A. E., Brown, C. T., Goffredi, S. K., Embaye, T. & Orphan, V. J. (2008) Diverse syntrophic partnerships from deep-sea methane vents revealed by direct cell capture and metagenomics. *Proceedings of the National Academy of Sciences of the United States of America*, **105**, 7052–7057.

Pinard, R., de Winter, A., Sarkis, G. J., et al. (2006) Assessment of whole genome amplification-induced bias through high-throughput, massively parallel whole genome sequencing. *BMC Genomics*, **7**, 216.

Prosser, J. I. (2010) Replicate or lie. *Environmental Microbiology*, **12**, 1806–1810.

Rinke, C., Schwientek, P., Sczyrba, A., et al. (2013) Insights into the phylogeny and coding potential of microbial dark matter. *Nature*, **499**, 431–437.

Rodriguez, R. L. & Konstantinidis, K. T. (2014) Estimating coverage in metagenomic data sets and why it matters. *ISME Journal*, **8**, 2349–2351.

Roux, S., Hawley, A. K., Torres Beltran, M., et al. (2014) Ecology and evolution of viruses infecting uncultivated SUP05 bacteria as revealed by single-cell and metagenomics. *Elife*, **3**, e03125.

Salter, S. J., Cox, M. J., Turek, E. M., et al. (2014) Reagent and laboratory contamination can critically impact sequence-based microbiome analyses. *BMC Biology*, **12**, 87.

Satinsky, B. M., Gifford, S. M., Crump, B. C. & Moran, M. A. (2013) Use of Internal standards for quantitative metatranscriptome and metagenome analysis. *Microbial Metagenomics, Metatranscriptomics, and Metaproteomics*, **531**, 237–250.

Satinsky, B. M., Crump, B. C., Smith, C. B., et al. (2014) Microspatial gene expression patterns in the Amazon River Plume. *Proceedings of the National Academy of Sciences of the United States of America*, **111**, 11085–11090.

Sboner, A., Mu, X. J., Greenbaum, D., Auerbach, R. K. & Gerstein, M. B. (2011) The real cost of sequencing: higher than you think! *Genome Biology*, **12**, 125.

Schirmer, M., Ijaz, U. Z., d'Amore, R., Hall, N., Sloan, W. T. & Quince, C. (2015) Insight into biases and sequencing errors for amplicon sequencing with the Illumina MiSeq platform. *Nucleic Acids Research*, **43**, e37.

Seemann, T. (2012) Tools to Merge Overlapping Paired-End Reads. Available at: http://thegenomefactory.blogspot.com/2012/11/tools-to-merge-overlapping-paired-end.html (accessed 30 October 2017).

Sharon, I. & Banfield, J. F. (2013) Genomes from metagenomics. *Science*, **342**, 1057–1058.

Sharon, I., Kertesz, M., Hug, L. A., et al. (2015) Accurate, multi-kb reads resolve complex populations and detect rare microorganisms. *Genome Research*, **25**, 534–543.

Stepanauskas, R. (2012) Single cell genomics: an individual look at microbes. *Current Opinion in Microbiology*, **15**, 613–620.

Swan, B. K., Martinez-Garcia, M., Preston, C. M., et al. (2011) Potential for chemo-lithoautotrophy among ubiquitous bacteria lineages in the dark ocean. *Science*, **333**, 1296–1300.

Tanner, M. A., Goebel, B. M., Dojka, M. A. & Pace, N. R. (1998) Specific ribosomal DNA sequences from diverse environmental settings correlate with experimental contaminants. *Applied and Environmental Microbiology*, **64**, 3110–3113.

Temperton, B. & Giovannoni, S. J. (2012) Metagenomics: microbial diversity through a scratched lens. *Current Opinion in Microbiology*, **15**, 605–612.

Thomas, T., Gilbert, J. & Meyer, F. (2012) Metagenomics – a guide from sampling to data analysis. *Microbial Informatics and Experimentation*, **2**, 3.

Thomas, T., Gilbert, J. & Meyer, F. (2015) A 123 of netagenomics. In: K. E. Nelson (ed.), *Genes, Genomes, and Metagenomes: Basics, Methods, Databases and Tools*. Springer, New York.

Tickle, T. L., Segata, N., Waldron, L., Weingart, U. & Huttenhower, C. (2013) Two-stage microbial community experimental design. *ISME Journal*, **7**, 2330–2339.

Tsementzi, D., Poretsky, R., Rodriguez, R. L., Luo, C. & Konstantinidis, K. T. (2014) Evaluation of metatranscriptomic protocols and application to the study of freshwater microbial communities. *Environmental Microbiology Reports*, **6**, 640–655.

Van Nieuwerburgh, F., Soetaert, S., Podshivalova, K., et al. (2011) Quantitative bias in Illumina TruSeq and a novel post amplification barcoding strategy for multi-plexed DNA and small RNA deep sequencing. *PLoS One*, **6**, e26969.

Venter, J. C., Remington, K., Heidelberg, J. F., et al. (2004) Environmental genome shotgun sequencing of the Sargasso Sea. *Science*, **304**, 66–74.

Weinstock, G. M. (2012) Genomic approaches to studying the human microbiota. *Nature*, **489**, 250–256.

Weiss, S., Amir, A., Hyde, E. R., Metcalf, J. L., Song, S. J. & Knight, R. (2014) Tracking down the sources of experimental contamination in microbiome studies. *Genome Biology*, **15**, 564.

Wendl, M. C., Kota, K., Weinstock, G. M. & Mitreva, M. (2013) Coverage theories for metagenomic DNA sequencing based on a generalization of Stevens' theorem. *Journal of Mathematical Biology*, **67**, 1141–1161.

Woyke, T., Xie, G., Copeland, A., et al. (2009) Assembling the marine metagenome, one cell at a time. *PLoS One*, **4**, e5299.

Xu, L., Brito, I. L., Alm, E. J. & Blainey, P. C. (2016) Virtual microfluidics for digital quantification and single-cell sequencing. *Nature Methods*, **13**, 759–762.

Yoon, H. S., Price, D. C., Stepanauskas, R., et al. (2011) Single-cell genomics reveals organismal interactions in uncultivated marine protists. *Science*, **332**, 714–717.

Zehr, J. P., Bench, S. R., Carter, B. J., et al. (2008) Globally distributed uncultivated oceanic N_2-fixing cyanobacteria lack oxygenic photosystem II. *Science*, **322**, 1110–1112.

Zhou, W., Chen, T., Zhao, H., et al.(2014) Bias from removing read duplication in ultra-deep sequencing experiments. *Bioinformatics*, **30**, 1073–1080.

5

Genomics of Single Species and Single Cells

Introduction

Traditionally, genome sequences were obtained from pure cultures of a single microbial species. More recently, methods for retrieving genome sequences from single cells and microbial communities have emerged. In all cases, DNA must be extracted and randomly fragmented into small pieces that are suitable for current DNA sequencing instruments prior to sequencing. Current DNA sequencing technologies (see section 4.3) produce sequence reads that are much shorter than microbial genomes (Fig. 5.1). Thus, in order to sequence a full genome, the short DNA sequences obtained from a DNA sequencing instrument must be joined together on the basis of overlapping sequences to form longer contiguous sequences called contigs (see Fig. 5.1). In turn, these contigs can be joined together into scaffolds using information about how pairs of DNA sequences are physically linked together on the same DNA fragment. Such genome assembly is not a trivial task, even for genomes of clonal cultures of a single species (heretofore referred to as isolate genomes) (Baker 2012).

This chapter provides a brief summary of the challenges of genome assembly and the solutions that have been produced so far. For more detailed information on genome assembly, the reader is referred to recent reviews on the topic (Compeau et al. 2011; Nagarajan & Pop 2013). Challenges and methods for metagenomic assembly are discussed later (see Chapter 6).

Genomic Approaches in Earth and Environmental Sciences, First Edition. Gregory Dick.
© 2019 John Wiley & Sons Ltd. Published 2019 by John Wiley & Sons Ltd.

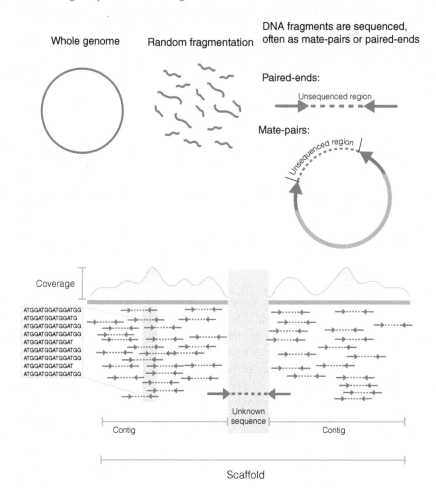

Figure 5.1 Schematic overview of genome sequencing. (*Top*) A genome is randomly fragmented and sequenced, often with sequenced reads that are paired-ends or mate-paired; these reads are physically linked, and this information is useful for subsequent assembly. Solid arrows represent sequence reads; dotted lines represent unsequenced regions from DNA fragments. (*Bottom*) Schematic representation of assembled contigs and scaffolds, showing overlapping sequences from contigs and paired-ends (*large bold arrows*) which are used to link contigs into scaffolds. Coverage, which is indicated above, is used to quantify the number of reads at each position within a contig or scaffold.

5.1 Algorithms for Genome Assembly

There are three main algorithms for *de novo* assembly: overlap-layout-consensus (OLC), de Bruijn graph, and string graph. OLC is an algorithm that was used by many early assemblers (Nagarajan & Pop 2013). It makes pairwise alignments between all reads, then merges reads into longer sequences based on overlaps. Pairwise alignment is computationally expensive, and the number of alignments scales as a square of the number of reads. Perhaps more importantly, therefore, this approach is generally not suited to the massive datasets of short reads that are now widely used (i.e., Illumina). The string graph approach, based on OLC, is more memory efficient but suffers from some of the same limitations as OLC. Recent reviews that provide more detail than the overview below are available (Baker 2012; Miller et al. 2010; Nagarajan & Pop 2013).

The de Bruijn graph method circumvents the need for pairwise alignments by breaking reads into shorter fragments called k-mers. From these k-mers, a de Bruijn graph is constructed, taking advantage of the fact that determining *exact matches* (linking k-mers) is much more computationally efficient than evaluating *similarity* (pairwise alignment). The de Bruijn graph is then used to reconstruct the genome using the computationally efficient Euler algorithm. An advantage of this approach is that graphs can be efficiently represented and explored (Compeau et al. 2011; Conway & Bromage 2011; Pell et al. 2012). De Bruijn graph approaches are currently the most popular and effective for large, short-read datasets, but they often require substantial computational resources in the form of random access memory (RAM). For an excellent tutorial on de Bruign graph approaches, see the homolog blog (www.homolog.us/Tutorials/index.php?p=6.2&s=1). For more details about de Bruijn graph approaches and their mathematical foundations, the reader is referred to Compeau et al. (2011) and references therein.

The advent of long-read technologies has challenged the field to develop new assembly approaches that take advantage of the ability of long reads to span repeats and other difficult-to-assemble regions while dealing with their higher error rates (see below). Original strategies were based on OCL, but a recent study showed the effectiveness and potential of adapting the de Bruijn graph approach for long error-prone reads (Lin et al. 2016). In addition to true *de novo* assembly, genome sequences from closely related organisms can also be used as templates to assist in assembly of new genomes (Gnerre et al. 2009).

5.2   Challenges of Genome Assembly

One of the major challenges of assembling microbial genomes is that they often contain sequences that are repeated at multiple locations of the genome. If DNA sequence reads produced by the DNA sequencer are too short to span such repeated regions and the unique sequences that flank them, a genome assembler cannot distinguish which location the repeat came from (Fig. 5.2). The result is that the repeats pile up, assembling together into one contig, which then has multiple conflicting paths into

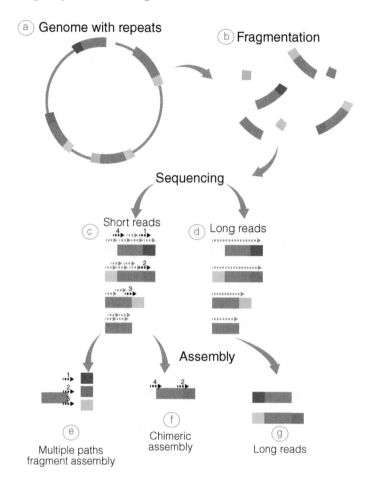

Figure 5.2 Challenges of genome assembly due to repeated sequence regions. (a) Schematic of a genome showing an identical repeat sequence region that is present in the genome in four copies (*red*). Unique sequence regions used further below are shown in other colors. For simplicity, the gray portion of the genome is not depicted in subsequent panels. (b) The genome is randomly fragmented for library preparation and sequencing. (c) Schematic of short reads (e.g., from Illumina), shown as dotted arrows in relation to genome fragments from (b). Key reads depicted below in (e) are shown in black and numbered. (d) Schematic of long reads (e.g., from PacBio), shown as dotted arrows in relation to genome fragments from (b). (e) Repeats sequenced with short read technology can result in disagreement between sequence reads that came from the repeat region at different genomic loci, leading to fragmentation of contigs. (f) Repeats sequenced with short-read technology can also result in chimeric assemblies in which genomic loci are erroneously brought together (yellow and brown in this case). (g) Long reads can resolve such repeat regions and thus are invaluable for producing accurate assemblies. Note that the various elements are not to scale.

adjacent genomic sequences. This can cause either termination of the contig or chimeric sequences, and it is particularly problematic for assembly of metagenomic data, where computational demand also becomes an issue (see Chapter 6).

5.3   Scaffolding

No matter how efficient or effective the assembly algorithm, in many circumstances the sequence data simply does not provide enough information to resolve repeats, close gaps, and reconstruct complete genomes. A key tool for resolving repeats and the order of contigs relative to each other is scaffolding, in which mate-pair or paired-end information is used to determine longer-range orientation of reads (see Fig. 5.1). Mate-pair and paired ends are similar in that both are sequence data from a pair of reads sequenced from the same DNA fragment. Whereas paired-ends are on the ends of a linear DNA fragment, mate-pairs are on a circularized fragment. These paired-ends and mate-pairs point towards each other and the distance between the reads is approximately known based on the genomic DNA fragmentation and size selection, providing valuable information that can be used during the assembly process, including the joining of contigs to form scaffolds.

Utilization of multiple DNA libraries with different insert sizes is a particularly effective method for assembling isolate genomes (Ribeiro et al. 2012). There are a number of tools for scaffolding (Hunt et al. 2014), and manual scaffolding is also typically possible for microbial genomes. Paired-end and mate-pair information, together with coverage, are also key for detecting assembly errors during the evaluation of assemblies (Fig. 5.3) (see section 5.5).

5.4   Programs and Pipelines for Genome Assembly

Numerous software packages are now available for assembly of microbial genome sequences (Table 5.1). Choosing an appropriate assembler depends on a number of factors, including which technology was used to produce the DNA sequence, the size and complexity of the dataset, and the performance of the assembler. Also important are practical considerations such as computational requirements (memory needs and software dependencies) and their scalability. Some assembly methods have been optimized for data derived from mixed communities and are discussed in the section on assembly of metagenomic data (see section 6.3).

For microbial genome sequences derived from pure cultures, Illumina sequencing is an effective and cost-efficient method of producing essentially complete draft genomes. Particularly effective are longer paired-end reads that can be merged to generate effective read lengths of greater than 400 bp. Recent open-source pipelines for assembling MiSeq data automate the entire process, including adapter trimming, quality filtering, error correction, generation of contigs and scaffolds, and detection of misassemblies (Coil et al. 2015; Tritt et al. 2012).

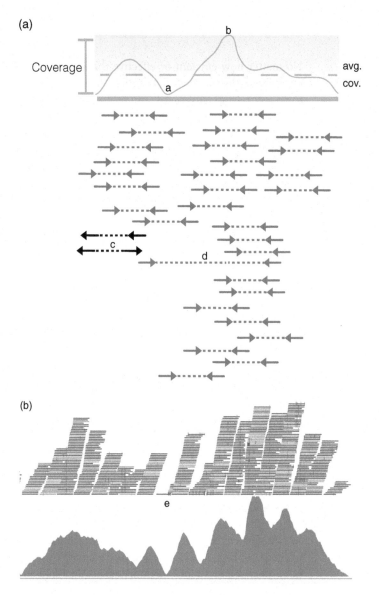

Figure 5.3 Identification of genome assembly problems using intrinsic features. (a) Schematic of a contig showing coverage and underlying paired-end reads as in Figure 5.1. a. A minimum in coverage can indicate a weak join. b. A peak in coverage well above the genome average may reflect erroneous assembly of a repeat region. c. Paired-end reads (*black*) that are improperly oriented indicate misassembly. d. Paired-end reads that are too far apart relative to the expected insert size. (b) Screenshot of read mapping (*top*) and average coverage (*bottom*). e. A weak join and likely chimera indicated by a coverage minimum and mismatched sequences between overlapping sequence reads.

Table 5.1 Methods for *de novo* genome assembly. Note that assemblers optimized for metagenomes are shown in Table 6.1.

Assembler	Method	Technology	Notes	References Original	Examples
Phrap	OLC	Sanger	Visualization provided by consed and phrapview	Green (1999)	Tyson et al. (2004)
Velvet	de Bruijn	Illumina, SOLiD, 454, Sanger		Zerbino and Birney (2008)	Anantharaman et al. (2016)
IDBA	de Bruijn	Illumina		Peng et al. (2010)	Smalley et al. (2015)
Mira	Greedy/OLC mix	Sanger, 454, Illumina, IonTorrent, PacBio	Flexible, high accuracy, well supported; hybrid capability	Chevreux et al. (1999)	Lesniewski et al. (2012)
ABySS	de Bruijn	Illumina		Simpson et al. (2009)	Wrighton et al. (2012)
A5	n/a	Illumina MiSeq	An integrated pipeline	Tritt et al. (2012); Coil et al. (2015)	Mansor et al. (2015)
Spades	de Bruijn	Illumina	Designed for single cell data; also works for multi-cell	Bankevich et al. (2012)	Lloyd et al. (2013)
ALLPATHS	de Bruijn	Illumina	Hybrid capability	Butler et al. (2008)	Farrer et al. (2009)
ABruijn	Combination of de Bruijn, OLC	Long reads (e.g., PacBio, Nanopore)	Designed for error-prone reads	Lin et al. (2016)	n/a

Source: Adapted from Nagarajan and Pop (2013).

While longer reads generated by Illumina MiSeq yield better assemblies than shorter reads, and can be used to finish smaller and simpler genomes, read lengths of the order of 500 bp are insufficient for resolving longer repeat sequence regions that are present in some larger and higher complexity genomes (Humbert et al. 2013). For such challenging genomes, the much longer reads provided by PacBio sequencing are key for obtaining complete genome sequences (Koren & Phillippy 2015). However, the lower accuracy of these long read technologies (see Table 4.1) is a challenge for genome assembly.

Several approaches have been used to circumvent the errors in long reads. First, combining error-prone long reads (e.g., PacBio) with high-accuracy short reads (e.g., Illumina) has proven to be effective (Koren et al. 2012; Utturkar et al. 2014). Although such a hybrid approach that takes advantage of the much cheaper Illumina sequencing may be more cost-efficient based on the sequencing itself, it requires at least two library preparations. A second approach uses only low-quality long reads by first assembling the long reads into highly accurate "preassembled reads" and then using those preassembled reads for the genome assembly (Chin et al. 2013). Third, new algorithms employ probabilistic approaches for the assembly of long reads

(Berlin et al. 2015). The costs of PacBio are now low enough (a complete genome, including library preparation and sequencing, can now be produced for <$500) that in many cases the streamlined approach provided by stand-alone PacBio sequencing is attractive.

5.5   Evaluation of Genome Assemblies

As illustrated by the above discussion, genome assemblies are prone to errors. It should be recognized that many assembled genomes in publicly available databases likely have errors; this is true even for genomes from pure cultures (Baker 2012). Assemblies of communities of organisms involve additional challenges, such as the risk of assembling sequences from similar yet distinct organisms (see Chapter 6). Plummeting sequencing costs have democratized genome sequencing to the point where small individual labs now sequence microbial genomes on a scale that was conducted only by large sequencing centers just a short time ago. Further, the increased throughput and decreased cost have come at the expense of read length (although the advent of high-accuracy PacBio data reverses that trend), which is a key determinant of genome assembly quality. Thus, more than ever, genome assemblies must be critically evaluated and interpreted with caution.

How can the accuracy of different assemblers be evaluated? How can errors in assemblies be detected? There are a variety of approaches.

First and most powerful, inherent characteristics of assemblies can be directly evaluated. For example, theoretically, genomic coverage should be relatively even around a genome, although some variability is to be expected based on stochastic effects and biases in library preparation and nucleotide composition. Therefore, stark discontinuities in coverage represent red flags that may indicate problems such as excess coverage due to repeats or insufficient coverage representing a weak join of reads in the contig (see Fig. 5.3). Misorientation of mate-pairs and paired-ends can also indicate a potential problem (see Fig. 5.3). Whereas assemblers such as Phrap and Mira have built-in capabilities for visualization, some assembly methods, especially those using de Bruijn graph approaches, do not directly keep track of reads with respect to contigs. Thus, reads must be mapped back to the contigs with methods like BWA (Li & Durbin 2009), then visualized with separate applications such as the Integrative Genomics Viewer (IGV) (Robinson et al. 2011) or Geneious (Kearse et al. 2012). IGV provides a guide to interpreting read pair orientations (BroadInstitute).

Second, there are tools to assess basic statistics of assemblies, such as the QUAST tool, which produces reports and summary tables of assembly results (Gurevich et al. 2013). Assembly metrics that do not require a reference genome are constrained to number and size of contigs, including the number of contigs, length of largest contig, total bases assembled, and *NX*, which is the length of the longest contig (L) such that all contigs of length $\geq L$ account for at least X% of the bases of the assembly. This is commonly reported as an N50. These metrics can be useful for quickly assessing the extent to which sequences are assembled into contigs, but they say nothing about the *quality* of the assembly in terms of misassemblies (there are more options for evaluating this where references genomes are present) (Gurevich et al. 2013).

Third, assemblies and assemblers can be evaluated by benchmarking them with datasets for which the answer is known. This can be done by using artificial datasets as in the genome assembly gold standard evaluations (GAGE) (Salzberg et al. 2012), assembleathon competitions (Bradnam et al. 2013), and *de novo* genome assembly project (dnGASP; http://cnag.bsc.es/). A concern with this approach is that artificial datasets may not accurately represent the data structure and issues of real genomes (Baker 2012). An alternative is to construct simulated datasets using real, high-quality, curated genomes (Junemann et al. 2014; Mavromatis et al. 2007; Utturkar et al. 2014). Additional methods for evaluating metagenomic assemblies are discussed below.

Another outcome of the transition towards cheaper sequencing and shorter read lengths is that the proportion of genomes that are only taken to the draft stage has increased dramatically (Chain et al. 2009). In the early days, sequencing was the major cost of a genome project, so it made sense to curate and close genomes. Now that sequencing represents a fraction of the cost, it is far more common to produce a draft genome that is never finished. This is not necessarily a problem; in some cases, draft genomes are sufficient for the scientific question at hand, and the cost and benefit of closing a genome (versus, for example, sequencing more genomes) are debatable (Fraser et al. 2002). For example, to determine the gene content of an organism, draft genomes are often sufficient provided that sequencing coverage is sufficiently high. However, major drawbacks of not closing genomes are (i) the inability to definitively say a gene is absent, and (ii) potential lack of information regarding ordering of contigs and hence overall genome organization and structure. In order to distinguish the quality of assembled genomes, there is a series of designations that reflect the degree of finishing and polishing that a genome has received (Chain et al. 2009) (Table 5.2). For the special case of viral genomes, separate categories and standards to convey assembly quality and utility have been put forward (Ladner et al. 2014).

5.6   Single-Cell Genomics

As discussed above and in Chapter 4, sometimes only a small subset of the community is of interest, and there are a variety of ways to target specific microbial cell types, guilds, or populations for sequencing. One of the most

Table 5.2 Designating the stage of genome finishing. Note that Parks et al. (2015) have proposed simplifying this scheme to *finished*, *noncontiguous finished*, and *draft*, with the quality of draft genomes being further quantitatively described by their completeness and contamination.

Designation	Meaning
Standard draft	Minimally or unfiltered data assembled into contigs. Can be relatively incomplete and likely has many regions of poor quality. This is the minimum standard for submission to public databases
High-quality draft	At least 90% complete. Efforts have been made to remove contaminating sequences, but little or no manual review of the product. No implied order and orientation to contigs. Sequence errors and misassemblies are possible. Appropriate for general assessment of gene content
Improved high-quality draft	Manual or automated improvements to the high-quality draft so that there are no discernible misassemblies. Some gaps have been closed to reduce the number of contigs and/or scaffolds. Low-quality regions and sequence errors may still be present. Suitable for comparative genomics
Annotation-directed improvement	In addition to the above, verification and correction of issues within coding regions such as frameshifts and stop codons. Repeat regions still not necessarily resolved. Valuable for gene comparisons and pathway reconstruction
Noncontiguous finished	High-quality assembly with automated and manual improvement. Closure approaches have been used to resolve most gaps, misassemblies, and low-quality regions. Only intractable gaps or repeats remain. Appropriate for most analyses
Finished	The gold standard; less than 1 error per 100 000 bp and single contiguous sequence. Fully manually reviewed and edited. Appropriate for all types of analysis

Source: Data from Chain et al. (2009).

powerful ways to conduct such targeted studies is by sequencing the genomes of individual microbial cells. There are now a variety of strategies and methods for single-cell genomics, and they have been detailed previously along with the strengths and weaknesses of this approach (Blainey 2013; Hedlund et al. 2014, 2015; Stepanauskas 2012; Xu et al. 2016). All single-cell approaches involve several steps: cellular isolation, lysis of the cell, amplification of the genome, and sequencing; each of these steps presents challenges that are discussed below following an overview of the strengths of single-cell genomics.

An obvious and major strength of single-cell genomics is that the "sampling" is discrete; individual cells are separated from other community members, including other closely related genotypes. Thus, some of the major challenges associated with metagenomic assembly (see Chapter 6), such as chimerism between different genotypes, are circumvented. Whereas metagenomic assemblies potentially represent composites of different genotypes, genomes derived from single cells can be assigned to single genotypes with high confidence. Because of its high-throughput capability, single-cell genomics can be used to survey the cell-to-cell genomic diversity within populations (Kashtan et al. 2014) and communities (Swan et al. 2011). In addition, the isolated cells carry associated viruses and endosymbionts, which are also sequenced during single-cell genomics. Thus, single-cell genomics is also a powerful way of linking cellular organisms with their associated viruses (Bhattacharya et al. 2013; Roux et al. 2014) and endosymbionts (Bhattacharya et al. 2012).

Single-cell genomics has advanced rapidly in the past 5 years but some difficult challenges remain. The amplification of genomic DNA, which is required to obtain enough material from one cell for sequencing, has stochastic biases that cause highly uneven coverage. The extent of this bias is such that some regions of the genome are highly covered while others are not sequenced at all, making it difficult to obtain complete genome sequences from single cells (though it can be done with finishing steps) (Woyke et al. 2010). Because the biased amplification is random, it can be resolved by sequencing multiple cells, but that can complicate the effort due to potentially different genotypes between cells, dulling one of the main strengths of single-cell genomics.

Separate from the issue of covering the whole genome, the uneven coverage also introduces problems for genome assembly (Nagarajan & Pop 2013), though recently developed assemblers accommodate such uneven coverage (Nurk et al. 2013; Peng et al. 2012). Another challenge of amplification of minute quantities of DNA is its susceptibility to contamination. This contamination can be mitigated and largely eliminated by clean lab procedures (de Bourcy et al. 2014) and downstream bioinformatic detection but the steps necessary to do so are costly, which is one of the reasons why single-cell genomics is done in just a few specialized labs (Hedlund et al. 2015).

The strengths and weaknesses of single cell genomics are largely complementary to those of metagenomics, such that these two approaches are synergistic (Hedlund et al. 2014; Lasken & McLean 2014).

References

Anantharaman, K., Brown, C. T., Burstein, D., et al. (2016) Analysis of five complete genome sequences for members of the class Peribacteria in the recently recognized Peregrinibacteria bacterial phylum. *PeerJ*, **4**, e1607.

Baker, M. (2012) De novo genome assembly: what every biologist should know. *Nature Methods*, **9**, 333–337.

Bankevich, A., Nurk, S., Antipov, D., et al. (2012) SPAdes: a new genome assembly algorithm and its applications to single-cell sequencing. *Journal of Computational Biology*, **19**, 455–477.

Berlin, K., Koren, S., Chin, C. S., Drake, J. P., Landolin, J. M. & Phillippy, A. M. (2015) Assembling large genomes with single-molecule sequencing and locality-sensitive hashing. *Nature Biotechnology*, **33**, 623–630.

Bhattacharya, D., Price, D. C., Yoon, H. S., et al. (2012) Single cell genome analysis supports a link between phagotrophy and primary plastid endosymbiosis. *Science Reports*, **2**, 356.

Bhattacharya, D., Price, D. C., Bicep, C., et al. (2013) Identification of a marine cyanophage in a protist single-cell metagenome assembly. *Journal of Phycology*, **49**, 207–212.

Blainey, P. C. (2013) The future is now: single-cell genomics of bacteria and archaea. *FEMS Microbiology Reviews*, **37**, 407–427.

Bradnam, K. R., Fass, J. N., Alexandrov, A., et al. (2013) Assemblathon 2: evaluating de novo methods of genome assembly in three vertebrate species. *Gigascience*, **2**, 10.

Broadinstitute. *Interpreting Color by Pair Orientation*. Available at: http://software.broadinstitute.org/software/igv/interpreting_pair_orientations (accessed 30 October 2017).

Butler, J., MacCallum, I., Kleber, M., et al. (2008) ALLPATHS: de novo assembly of whole-genome shotgun microreads. *Genome Research*, **18**, 810–820.

Chain, P. S., Grafham, D. V., Fulton, R. S., et al. (2009) Genomics. Genome project standards in a new era of sequencing. *Science*, **326**, 236–237.

Chevreux, B., Wetter, T. & Suhai, S. (1999) *Genome Sequence Assembly Using Trace Signals and Additional Sequence Information*. Available at: www.bioinfo.de/isb/gcb99/talks/chevreux/ (accessed 30 October 2017).

Chin, C. S., Alexander, D. H., Marks, P., et al. (2013) Nonhybrid, finished microbial genome assemblies from long–read SMRT sequencing data. *Nature Methods*, **10**, 563–569.

Coil, D., Jospin, G. & Darling, A. E. (2015) A5-miseq: an updated pipeline to assemble microbial genomes from Illumina MiSeq data. *Bioinformatics*, **31**, 587–589.

Compeau, P. E., Pevzner, P. A. & Tesler, G. (2011) How to apply de Bruijn graphs to genome assembly. *Nature Biotechnology*, **29**, 987–991.

Conway, T. C. & Bromage, A. J. (2011) Succinct data structures for assembling large genomes. *Bioinformatics*, **27**, 479–486.

De Bourcy, C. F., de Vlaminck, I., Kanbar, J. N., Wang, J., Gawad, C. & Quake, S. R. (2014) A quantitative comparison of single-cell whole genome amplification methods. *PLoS One*, **9**, e105585.

Farrer, R. A., Kemen, E., Jones, J. D. & Studholme, D. J. (2009) De novo assembly of the Pseudomonas syringae pv. syringae B728a genome using Illumina/Solexa short sequence reads. *FEMS Microbiology Letters*, **291**, 103–111.

Fraser, K. P. P., Clarke, A. & Peck, L. S. (2002) Feast and famine in Antarctica: seasonal physiology in the limpet Nacella concinna. *Marine Ecology Progress Series*, **242**, 169–177.

Gnerre, S., Lander, E. S., Lindblad-Toh, K. & Jaffe, D. B. (2009) Assisted assembly: how to improve a de novo genome assembly by using related species. *Genome Biology*, **10**, R88.

Green, P. (1999) *Documentation for Phrap and CrossMatch (version 0.990319)*. Available at: www.phrap.org/phredphrapconsed.html (accessed 30 October 2017).

Gurevich, A., Saveliev, V., Vyahhi, N. & Tesler, G. (2013) QUAST: quality assessment tool for genome assemblies. *Bioinformatics*, **29**, 1072–1075.

Hedlund, B. P., Dodsworth, J. A., Murugapiran, S. K., Rinke, C. & Woyke, T. (2014) Impact of single-cell genomics and metagenomics on the emerging view of extremophile "microbial dark matter". *Extremophiles*, **18**, 865–875.

Hedlund, B. P., Dodsworth, J. A. & Staley, J. T. (2015) The changing landscape of microbial biodiversity exploration and its implications for systematics. *Systematic and Applied Microbiology*, **38**, 231–236.

Humbert, J. F., Barbe, V., Latifi, A., et al. (2013) A tribute to disorder in the genome of the bloom-forming freshwater cyanobacterium Microcystis aeruginosa. *PLoS One*, **8**, e70747.

Hunt, M., Newbold, C., Berriman, M. & Otto, T. D. (2014) A comprehensive evaluation of assembly scaffolding tools. *Genome Biology*, **15**, R42.

Junemann, S., Prior, K., Albersmeier, A., et al. (2014) GABenchToB: a genome assembly benchmark tuned on bacteria and benchtop sequencers. *PLoS One*, **9**, e107014.

Kashtan, N., Roggensack, S. E., Rodrigue, S., et al. (2014) Single-cell genomics reveals hundreds of coexisting subpopulations in wild Prochlorococcus. *Science*, **344**, 416–420.

Kearse, M., Moir, R., Wilson, A., et al. (2012) Geneious Basic: an integrated and extendable desktop software platform for the organization and analysis of sequence data. *Bioinformatics*, **28**, 1647–1649.

Koren, S. & Phillippy, A. M. (2015) One chromosome, one contig: complete microbial genomes from long-read sequencing and assembly. *Current Opinion in Microbiology*, **23**, 110–120.

Koren, S., Schatz, M. C., Walenz, B. P., et al. (2012) Hybrid error correction and de novo assembly of single-molecule sequencing reads. *Nature Biotechnology*, **30**, 693–700.

Ladner, J. T., Beitzel, B., Chain, P. S., et al. (2014) Standards for sequencing viral genomes in the era of high-throughput sequencing. *MBio*, **5**, e01360-14.

Lasken, R. S. & Mclean, J. S. (2014) Recent advances in genomic DNA sequencing of microbial species from single cells. *Nature Reviews Genetics*, **15**, 577–584.

Lesniewski, R. A., Jain, S., Anantharaman, K., Schloss, P. D. & Dick, G. J. (2012) The metatranscriptome of a deep-sea hydrothermal plume is dominated by water column methanotrophs and lithotrophs. *ISME Journal*, **6**, 2257–2268.

Li, H. & Durbin, R. (2009) Fast and accurate short read alignment with Burrows–Wheeler transform. *Bioinformatics*, **25**, 1754–1760.

Lin, Y., Yuan, J., Kolmogorov, M., Shen, M. W., Chaisson, M. & Pevzner, P. A. (2016) Assembly of long error-prone reads using de Bruijn graphs. *Proceedings of the National Academy of Sciences of the United States of America*, **113**, E8396–E8405.

Lloyd, K. G., Schreiber, L., Petersen, D. G., et al. (2013) Predominant archaea in marine sediments degrade detrital proteins. *Nature*, **496**, 215–218.

Mansor, M., Hamilton, T. L., Fantle, M. S. & Macalady, J. L. (2015) Metabolic diversity and ecological niches of Achromatium populations revealed with single-cell genomic sequencing. *Frontiers in Microbiology*, **6**, 822.

Mavromatis, K., Ivanova, N., Barry, K., et al. (2007) Use of simulated data sets to evaluate the fidelity of metagenomic processing methods. *Nature Methods*, **4**, 495–500.

Miller, J. R., Koren, S. & Sutton, G. (2010) Assembly algorithms for next-generation sequencing data. *Genomics*, **95**, 315–327.

Nagarajan, N. & Pop, M. (2013) Sequence assembly demystified. *Nature Reviews Genetics*, **14**, 157–167.

Nurk, S., Bankevich, A., Antipov, D., et al. (2013) Assembling single-cell genomes and mini-metagenomes from chimeric MDA products. *Journal of Computational Biology*, **20**, 714–737.

Parks, D. H., Imelfort, M., Skennerton, C. T., Hugenholtz, P. & Tyson, G. W. (2015) CheckM: assessing the quality of microbial genomes recovered from isolates, single cells, and metagenomes. *Genome Research,* **25**, 1043–1055.

Pell, J., Hintze, A., Canino-Koning, R., Howe, A., Tiedje, J. M. & Brown, C. T. (2012) Scaling metagenome sequence assembly with probabilistic de Bruijn graphs. *Proceedings of the National Academy of Sciences of the United States of America*, **109**, 13272–13277.

Peng, Y., Leung, H. C. M., Yiu, S. M. & Chin, F. Y. L. (2010) *IDBA – A Practical Iterative De Bruijn graph De Novo Assembler*. 14th Annual International Conference on Research in Computational Molecular Biology (RECOMB 2010), Lisbon, Portugal. Springer, pp. 426–440.

Peng, Y., Leung, H. C., Yiu, S. M. & Chin, F. Y. (2012) IDBA-UD: a de novo assembler for single-cell and metagenomic sequencing data with highly uneven depth. *Bioinformatics*, **28**, 1420–1428.

Ribeiro, F. J., Przybylski, D., Yin, S., et al. (2012) Finished bacterial genomes from shotgun sequence data. *Genome Research*, **22**, 2270–2277.

Robinson, J. T., Thorvaldsdottir, H., Winckler, W., et al. (2011) Integrative genomics viewer. *Nature Biotechnology*, **29**, 24–26.

Roux, S., Hawley, A. K., Torres Beltran, M., et al. (2014) Ecology and evolution of viruses infecting uncultivated SUP05 bacteria as revealed by single-cell and metagenomics. *Elife*, **3**, e03125.

Salzberg, S. L., Phillippy, A. M., Zimin, A., et al. (2012) GAGE: a critical evaluation of genome assemblies and assembly algorithms. *Genome Research*, **22**, 557–567.

Simpson, J. T., Wong, K., Jackman, S. D., Schein, J. E., Jones, S. J. & Birol, I. (2009) ABySS: a parallel assembler for short read sequence data. *Genome Research*, **19**, 1117–1123.

Smalley, M. D., Marinov, G. K., Bertani, L. E. & Desalvo, G. (2015) Genome sequence of Magnetospirillum magnetotacticum strain MS-1. *Genome Announcements*, **3**, ii.

Stepanauskas, R. (2012) Single cell genomics: an individual look at microbes. *Current Opinion in Microbiology*, **15**, 613–620.

Swan, B. K., Martinez-Garcia, M., Preston, C. M., et al. (2011) Potential for chemolithoautotrophy among ubiquitous bacteria lineages in the dark ocean. *Science*, **333**, 1296–1300.

Tritt, A., Eisen, J. A., Facciotti, M. T. & Darling, A. E. (2012) An integrated pipeline for de novo assembly of microbial genomes. *PLoS One*, **7**, e42304.

Tyson, G. W., Chapman, J., Hugenholtz, P., (2004) Community structure and metabolism through reconstruction of microbial genomes from the environment. *Nature*, **428**, 37–43.

Utturkar, S. M., Klingeman, D. M., Land, M. L., et al. (2014) Evaluation and validation of de novo and hybrid assembly techniques to derive high-quality genome sequences. *Bioinformatics*, **30**, 2709–2716.

Woyke, T., Tighe, D., Mavromatis, K., et al. (2010) One bacterial cell, one complete genome. *PLoS One*, **5**, e10314.

Wrighton, K. C., Thomas, B. C., Miller, C. S., et al. (2012) Fermentation, hydrogen, and sulfur metabolism in multiple uncultivated bacterial phyla. *Science*, **337**, 1661–1665.

Xu, L., Brito, I. L., Alm, E. J. & Blainey, P. C. (2016) Virtual microfluidics for digital quantification and single-cell sequencing. *Nature Methods*, **13**, 759–762.

Zerbino, D. R. & Birney, E. (2008) Velvet: algorithms for de novo short read assembly using de Bruijn graphs. *Genome Research*, **18**, 821–829.

Metagenomics

Assembly and Database-Dependent Approaches

Introduction

The DNA sequence of whole genomes obtained directly from assemblages of multiple microbial taxa can provide powerful insights into the structure of microbial communities and their interaction with geochemical and environmental processes. This overall approach, called "metagenomics," "community genomics" or "environmental genomics," can be conducted by many different methods of sampling processing and data analysis. Considerations for sampling and experimental design and overall approach are common to other omics approaches and are described in Chapter 4. Here, we focus on strategies for analyzing metagenomic data. A key decision is whether to analyze it at the level of DNA sequence reads or to assemble the reads into contigs, scaffolds, and genomic bins. Although the decision on whether to assemble or not depends on the scientific question at hand, there are some overwhelming advantages to assembling reads into contigs and genomes, and a key hindrance to database-dependent approaches is the lingering lack of sufficient reference genomes that represent the genetic diversity inherent in natural microbial communities.

6.1 To Assemble or Not To Assemble?

Following sequencing, there are several different ways to approach the analysis of shotgun metagenomic dataset (Fig. 6.1). One of the first and most important questions is whether to assemble the reads into larger contigs or to conduct the analysis at the read level, either by directly annotating reads or by mapping them to reference genomes. A number of issues are factored

Genomic Approaches in Earth and Environmental Sciences, First Edition. Gregory Dick.
© 2019 John Wiley & Sons Ltd. Published 2019 by John Wiley & Sons Ltd.

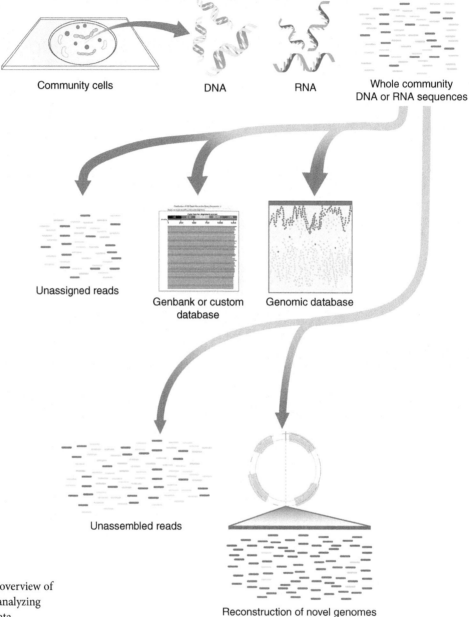

Community cells

DNA

RNA

Whole community
DNA or RNA sequences

Unassigned reads

Genbank or custom
database

Genomic database

Unassembled reads

Reconstruction of novel genomes

Figure 6.1 An overview of approaches for analyzing metagenomic data.

into this decision. First, the scientific question should dictate the best approach. For example, evaluation of the diversity and abundance of a modest set of well-known genes or functions that have been curated with custom databases may be effectively conducted at the read level (Reed et al. 2015). Broader functional profiling of whole communities may also be conducted at the read level, though the resolution of such studies is somewhat coarse.

On the other hand, assembly provides crucial information about which functions belong to which organisms (Anantharaman et al. 2016a; Hultman et al. 2015; Lesniewski et al. 2012; Sheik et al. 2014; Tyson et al. 2004; Venter et al. 2004; Wrighton et al. 2012), insights into the evolutionary processes in natural populations (Allen & Banfield 2005; Allen et al. 2007; Simmons et al. 2008) and the form, tempo, and ecological implications of strain-level variation (Andersson & Banfield 2008; Denef & Banfield 2012; Denef et al. 2010a, 2010b; Lo et al. 2007). Some studies are focused on a particular species for which there are many reference genomes available, so it is appropriate and feasible to use a reference genome-based approach. However, there remain many entire clades and even divisions without reference genome sequences (Baker & Dick 2013), despite laudable efforts such as the GEBA project to enhance phylogenetic coverage of genome sequences (Wu et al. 2009). Further, the extensive genomic diversity within microbial species means that reference genomes will often miss genes that encode key local environmental adaptations. The limitations of using reference genomes should be kept in mind when designing strategies for analysis of metagenomic data, and *de novo* assembly of genomes from metagenomes can fill this gap (Howe & Chain 2015).

A second major consideration in choosing the best approach to analyzing metagenomic data is the feasibility of assembly given the target community. In the early days of metagenomics, only the lowest diversity microbial communities yielded substantial assemblies (Tyson et al. 2004). With today's massively high-throughput DNA sequencing technologies, high-diversity communities can now yield thousands of essentially complete genomes from diverse microbial communities (Anantharaman et al. 2016b). However, some of the highest diversity communities, such as those in soil, are still refractory to assembly. Assembly-based approaches to such communities may incorporate only a small fraction of the reads, although improved assembly methods are leading to advances in this area (Crusoe et al. 2015; Howe et al. 2014).

Finally, there are several practical advantages to assembly, even if downstream analysis will be at the gene level. Longer contigs that have multiple reads covering each position enable correction of sequencing errors and more accurate annotation methods than do short reads (Thomas et al. 2012). Assembly also compresses datasets into more manageable and computationally tractable forms, though this may also be accomplished by clustering methods.

6.2 Database-Dependent Approaches

As discussed above, because of the incredible diversity of microbial communities, both in terms of novel organisms that have not been sequenced and highly variable genomic content within microbial species, the utility of reference databases for interpretation of metagenomic data is limited. However, leveraging the prior knowledge available in databases can be effective, especially for organisms and functions that are well known.

There are three main approaches to analyzing metagenomic databases by comparison to databases:

- by comparing reads to publicly available or custom databases with programs such as BLAST (Altschul et al. 1990) or DIAMOND (Buchfink et al. 2015)
- via large publicly available pipelines such as MG-RAST that perform taxonomic and functional information at the read level
- by mapping reads to reference genomes by a process called fragment recruitment.

This section provides an overview of these approaches, whereas subsequent sections focus on specific methods of functional and taxonomic annotations (see Chapter 8).

Direct comparison of sequence reads to databases with BLAST was one of the first approaches used to analyze metagenomic and metatranscriptomic data, and it remains popular due to its flexibility. Comparisons can be conducted against databases ranging from the whole NCBI nonredundant database (NR) (DeLong et al. 2006) and RefSeq (Satinsky et al. 2015) to custom databases containing special functions of interest (Reed et al. 2015). For whole-community overviews and comparative analysis, many researchers find that using more highly curated databases, in which sequences are organized into hierarchical functional categories or metabolic pathways, is useful; these include COG (Tatusov et al. 2000), KEGG (Kanehisa et al. 2016), SEED (Overbeek et al. 2005), and MetaCyc (Caspi et al. 2016). However, this higher degree of curation and organization comes at the expense of losing sequences in the database that may be critical for some queries and applications.

A major challenge with using large databases such as NR is the computational time – enormous computational resources and/or compute time are required. For all BLAST-based approaches, a weakness is the "top-hit" effect, wherein the best match is not necessarily the right answer because the difference between the query and subject may be large and the difference between the best hit and the second best hit may be small. A related issue is the difficulty in choosing a similarity threshold for positive matches, because the relationship between conservation of sequence and function varies widely across different protein families. Programs such as MEGAN (Huson et al. 2007) and Darkhorse (Podell & Gaasterland 2007), that consider a number of hits for each query and the extent which they agree with each other, enable more accurate annotations and/or generation of confidence scores and help to alleviate this issue. Methods that use calculated gene- and position-specific thresholds are also critical for maximizing the accuracy of analyses of short-read datasets (Orellana et al. 2017).

The metagenomics RAST (MG-RAST) server is perhaps the most widely used tool for analyzing metagenomic sequences at the sequence read level. MG-RAST is an automated and integrated pipeline for analyzing metagenomic data, and includes methods not just for annotating single reads but also for analysis of assembled contigs, upstream quality control and downstream comparison of datasets, statistical analysis and visualization, as well as tools for import and export of data (Meyer et al. 2008). However, its read-level capabilities distinguish it from other platforms. Although MG-RAST is the best publicly available resource of its kind, it suffers from several issues inherent to read level analysis and generalized annotation. Functional annotations are often quite coarse, being limited in their specificity by the short length of single reads (~100 bp). Some key functions of interest, such as certain sulfur cycling genes, are not included in databases used by MG-RAST (Crowe et al. 2014). Finally, as a free service for computationally intensive processing of enormous datasets, it is no surprise that the turnaround time for annotating datasets by MG-RAST can be months.

A third method of leveraging reference genomes or scaffolds to analyze metagenomic data at the read level is fragment recruitment, in which reads from environmental populations are mapped onto reference genomes (Rusch et al. 2007). The mapping is typically conducted at the nucleotide level with BLASTn (or k-mer-based mappers such as Bowtie (Langmead & Salzberg 2012) or BWA (Li & Durbin 2009) where only close matches are of interest), and the output is visualized as the percent identity of each read plotted by its location on the reference genome (Fig. 6.2). The results provide information about the presence, abundance, and genomic content of related populations. Reads that plot at high nucleotide identify (95–100%) are likely from populations closely related to the reference genome, whereas others (70–90%) stem from related but distinct populations, and these results can be interpreted in the context of average nucleotide identity for phylogenetic and taxonomic purposes (Rodriguez-R & Konstantinidis 2014). Highly conserved regions of the genome and genomic islands present in the reference genome but not in the wild populations are easily identified (see Fig. 6.2). Differences in gene order can also be identified through the orientation of mate-pair or paired-end reads (Rusch et al. 2007) (see also Fig. 5.3).

Although fragment recruitment can be a powerful method of exploring how wild populations compare to reference genomes, it is important to recognize that genomic islands present in the wild populations but not in the reference genome will be missed. Since such variable portions of the genome often encode key adapations to the local environment, this is a critical weakness of reference genome and database-dependent approaches. These same capabilities and limitations also apply to the use of reference genomes for analysis of metatranscriptomic and metaproteomics data. Combining the use of these several database-dependent strategies offers valuable complementary perspectives on metagenomic and metatranscriptomic datasets (Shi et al. 2011).

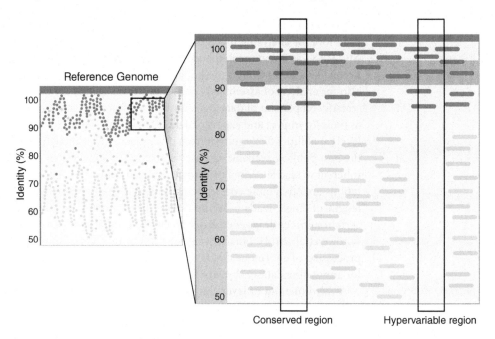

Figure 6.2 A schematic example of fragment recruitment. Reads from a metagenome (or transcriptome) are plotted against a reference genome according to their percent identity. It is often the case that one population in the environment will have high similarity to the reference genome and several others will be more divergent, with a natural "gap" in sequence space in between.

6.3 Database-Independent Approaches: *De Novo* Assembly

De novo assembly is highly advantageous because it can potentially capture heterogeneity that is inherent to natural populations of microorganisms. However, assembly of metagenomic data presents special challenges above and beyond those of assembling genomes from single organisms (see section 5.2) (Howe & Chain 2015; Pop 2009; Wooley et al. 2010).

First, there is the issue of extensive taxonomic diversity of microbial communities. Because different taxa have large differences in gene sequences and content, even among closely related organisms, there are typically staggering numbers of genotypes present in microbial communities, and they must be accounted for when calculating sequencing effort required for desired genomic coverage. This problem is particularly acute when low-abundance organisms are of interest; extensive sequencing must be done to sequence through the abundant members and achieve sufficient coverage of low-abundance members. As a result, high-complexity microbial communities require enormous volumes of sequence data, producing large datasets that are challenging in terms of the computer memory needed for assembly. Continued improvements in scalability of metagenomic assembly are required to tackle this issue (Boisvert et al. 2012; Crusoe et al. 2015; Howe et al. 2014; Li et al. 2015; Muggli et al. 2017; Pell et al. 2012).

Second, microbial communities seldom contain discrete populations that are completely distinct from each other at the genomic level. Rather, within taxonomic groups, there are typically multiple closely related genotypes in the same environment that share some genes but also contain genes that are unique to each genotype (Allen & Banfield 2005; Anantharaman et al. 2013; Sharon and Banfield 2013) (see section 2.4). The shared genes vary in their levels of sequence similarity between genotypes; some genes may even be conserved at 100% sequence identity while others show substantial dissimilarity. The conserved genes may co-assemble, leading to erroneous chimeric sequences (Fig. 6.3). This can be particularly troublesome for highly conserved genes (e.g., rRNA genes) or mobile elements (plasmids or transposases) that are identical between different strains or even species. Depending on the level of dissimilarity and the assembly platform and parameters used, different alleles of the same genes may be co-assembled into the same contig or segregated into different contigs. The former leads to composite genome assemblies containing sequence diversity. Further, related genotypes often display differences in gene order and orientation (Allen & Banfield 2005), leading to multiple paths in metagenomic assemblies, which may cause premature termination of contigs and even lack of assembly altogether (see Fig. 6.3). For de Bruijn graph assemblies of short read data, the extent of this problem is such that even dominant organisms can be missed when there are multiple closely related organisms present (Sharon et al. 2015). Paired-end and mate-pair information can be used to resolve such issues in some cases (Iverson et al. 2012). Coverage is also a key source of information for separating contigs from closely related organisms (see Chapter 7).

Finally, a third challenge of metagenomic assembly is that different organisms in a community are present at varying abundances and will therefore have different genomic coverage, which is typically flagged as a problem by traditional isolate genome assemblers. Most currently available genome assemblers were designed for assembly of genomes from single organisms rather than communities of organisms. However, there are now several assemblers that address this and other major challenges associated with assembly of metagenomic data (Table 6.1). A more detailed overview of current assemblers is provided by Vollmers et al. (2017).

IDBA-UD introduced several new strategies for dealing with metagenomic data and compares favorably with other assemblers in its performance in assembling metagenomic data (Peng et al. 2012). In addition to using multiple k-mers as in IDBA (Peng et al. 2010), it uses a more sophisticated method for addressing uneven coverage and paired-end information to resolve local repeat sequences that otherwise introduce branches into the de Bruijn graph. IDBA-UD is a favorite of the author's lab and has been employed with high success to complex communities, including groundwater (Brown et al. 2015), deep-sea hydrothermal plumes (Anantharaman

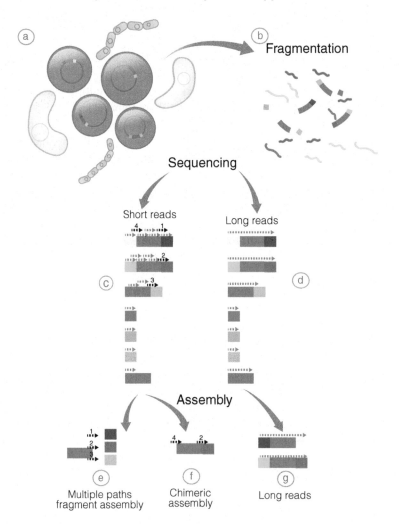

Figure 6.3 Pitfalls of metagenomic assembly. (a) Schematic of microbial cells of three different species and genomes. Here we focus on the red species with circular cells, in which the thicker blocks of the genome represent identical sequence repeats between different strains. (b) The genomes are randomly fragmented for library preparation and sequencing. (c) Schematic of short reads sequencing technology (e.g., Illumina). Sequence reads are shown as dotted arrows in relation to genome fragments from (b). Key reads depicted below in (e) are shown in black and numbered. (d) Schematic of long read sequencing technology (e.g., PacBio). Sequence reads are shown as dotted arrows in relation to genome fragments from (b). (e) Repeats sequenced with short read technology can result in disagreement between sequence reads that came from the repeat region at different genomic loci, leading to fragmentation of contigs. (f) Repeats sequenced with short read technology can also result in chimeric assemblies in which genomic loci are erroneously brought together (yellow and brown in this case). (g) Long reads can resolve such repeat regions and thus are invaluable for producing accurate assemblies. Note that the various elements are not to scale.

Table 6.1 Selected sequence assemblers for metagenomic data. Note that assemblers not optimized for metagenomic data are shown in Table 5.1.

Assembler	Method	Technology	Notes	References
Metavelvet	de Bruijn	Illumina, SOLiD, 454, Sanger	Uses coverage and connectivity to decompose the de Bruijn graph	Namiki et al. 2012
Meta-IDBA	de Bruijn	Illumina		Peng et al. 2011; Howe et al. 2014
IDBA-UD	de Bruijn	Illumina	Also for single-cell genomes	Peng et al. 2012
MEGA-HIT	de Bruijn	Illumina	Uses some principles of IDBA, SOAP-de novo	Li et al. 2015
Ray Meta	de Bruijn	Illumina	Scalable across nodes by message passing	Boisvert et al. 2012
Omega	String graph ("overlap graph")		Good for MiSeq	Haider et al. 2014

Source: Adapted from Nagarajan and Pop (2013).

et al. 2016a), and estuary sediments (Baker et al. 2015). An interesting aspect of assembly that seems particularly acute with IDBA-UD is that it performs best at low to intermediate coverage. At higher coverage, contigs are fragmented, so performing assembly with a subset of reads may be optimal (Handley et al. 2014; Hug et al. 2016).

The new assembler MEGAHIT (Li et al. 2015) combines some features of IDBA-UD (namely the multiple k-mer size strategy) with the computational benefits of succinct de Bruijn graphs (Bowe et al. 2012) to produce a metagenomic assembler that is reportedly fast and accurate. MetaVelvet built on the genome assembler Velvet by using coverage and graph connectivity information to decompose the metagenomic de Bruijn graph into subgraphs representing single species (Namiki et al. 2012).

The advent of longer reads promises to once again change the landscape of assembly methods and outcomes for metagenomics. Improved assembly of MiSeq reads (300 bp) has been achieved with string overlap methods (Haider et al. 2014). PacBio offers read lengths 10 kb and longer; although more expensive than Illumina, this technology is already being used routinely for sequencing of isolate genomes because of the assembly advantage conferred by longer reads (see Chapter 5). For metagenomic data, where cost efficiency and throughput are at a premium, the optimal approach currently may be to conduct a combination of short and long read sequencing. This approach has been shown to effectively reconstruct genomes and, intriguingly, seems to be particularly powerful in resolving lower-abundance community members (Sharon et al. 2015).

In addition to improvements in algorithms for metagenomic assembly, advances in methods upstream and downstream of assembly show promise. For example, physical cross-linking of DNA can be used to produce "contact probability maps" that help discern which reads came from the same individual cell within a community (Beitel et al. 2014; Burton et al. 2014), though this method has not yet been widely applied. Preassembly processing such as digital normalization and k-mer partitioning also shows promise (Cleary et al. 2015; Crusoe et al. 2015; Howe et al. 2014). Postassembly processes for merging contigs can also improve assembly outcomes (Scholz et al. 2014).

6.4 Evaluation of Metagenomic Assemblies

Methods for evaluating metagenomic assemblies are similar to the evaluation of assemblies from isolate genomes, including inherent characteristics such as coverage and mate-pair/paired-end integrity. Basic statistics on real datasets have also been used to evaluate metagenomic assemblers (Vollmers et al. 2017). In addition, binning can be used to identify chimeric contigs, which are one of the major challenges with metagenomic assemblies. Contigs containing regions that fall into different genomic bins are highly likely to be chimeric (or, less likely, to contain horizontally transferred genes, which may retain the genomic signature of the donor organism) (Lawrence & Ochman 1998).

Finally, benchmarking of metagenomic assemblers had been done by using standardized datasets in which the solution is known. Mavromatis et al. (2007) constructed simulated datasets of varying complexity by combining sequencing reads randomly selected from 113 isolate genomes. The reads were assembled and binned and evaluated relative to the "answer key." Shakya and colleagues also present a synthetic metagenomic dataset (Shakya et al. 2013). This is a powerful method for understanding the limitations and challenges of assembly, but it is imperfect in that simulated datasets may not accurately represent the various forms of diversity that are present within real microbial communities. An alternative is to spike in reference genomes into real metagenomes (Luo et al. 2012). This paper found that while single genotypes can be effectively reconstructed as long as genomic coverage is sufficiently high (>20×), at lower coverage there are problems with chimeras. In both cases, closely related genotypes could not be resolved.

6.5 A Philosophy of Metagenome Assemblies

Now that we have toured the ins and outs of metagenomic assembly, and considering what we know about microbial genomic diversity, how should we view genome assemblies? They almost always have errors, even for "clonal" genomes from pure cultures. As we discussed above, metagenomes have many pitfalls, and even when chimeras are not present, genomes assembled from metagenomes are composites from numerous cells and thus likely do not reflect the exact genetic make-up of any single cell. However, this does not mean that metagenomic assemblies should be viewed as right or wrong. Assemblies can be viewed as constructs, as abstractions of the data and as a framework for interpreting and visualizing genomes and metagenomes and the genomic heterogeneity that is inherent to microbial populations and communities. An important conclusion is that the consensus sequence represents just that – a composite that potentially masks underlying variation. In fact, assemblies can be an effective means of studying this underlying diversity. For example, mate-pair or paired-end data can be used to scaffold and to identify gaps, insertions, rearrangements, etc. (Allen & Banfield 2005). SNP patterns can reveal population variation and evolutionary processes (Allen et al. 2007; Andersson & Banfield 2008; Rosen et al. 2015).

Finally, the complexities and pitfalls of metagenomic assembly must be weighed against the limitations of other approaches. Given extensive microbial genomic diversity, we should be wary of using reference genomes because they may be missing key aspects of the genomes being studied. Single-cell genomics offers great promise for (ever-increasing) high-throughput exploration of genomic diversity (Kashtan et al. 2014), but it lacks the wide-angle, whole-community view offered by metagenomics, and single-cell genome assemblies offer their own challenges.

References

Allen, E. E. & Banfield, J. F. (2005) Community genomics in microbial ecology and evolution. *Nature Reviews Microbiology*, **3**, 489–498.

Allen, E. E., Tyson, G. W., Whitaker, R. J., Detter, J. C., Richardson, P. M. & Banfield, J. F. (2007) Genome dynamics in a natural archaeal population. *Proceedings of the National Academy of Sciences of the United States of America*, **104**, 1883–1888.

Altschul, S. F., Gish, W., Miller, W., Myers, E. W. & Lipman, D. J. (1990) Basic local alignment search tool. *Journal of Molecular Biology*, **215**, 403–410.

Anantharaman, K., Breier, J. A., Sheik, C. S. & Dick, G. J. (2013) Evidence for hydrogen oxidation and metabolic plasticity in widespread deep-sea sulfur-oxidizing bacteria. *Proceedings of the National Academy of Sciences of the United States of America*, **110**, 330–335.

Anantharaman, K., Breier, J. A. & Dick, G. J. (2016a) Metagenomic resolution of microbial functions in deep-sea hydrothermal plumes across the Eastern Lau Spreading Center. *ISME Journal*, **10**, 225–239.

Anantharaman, K., Brown, C. T., Hug, L. A., et al. (2016b) Thousands of microbial genomes shed light on interconnected biogeochemical processes in an aquifer system. *Nature Communications*, **7**, 13219.

Andersson, A. F. & Banfield, J. F. (2008) Virus population dynamics and acquired virus resistance in natural microbial communities. *Science*, **320**, 1047–1050.

Baker, B. J. & Dick, G. J. (2013) Omic approaches in microbial ecology: charting the unknown. *Microbe*, **8**, 353–360.

Baker, B. J., Lazar, C., Teske, A. & Dick, G. J. (2015) Genomic resolution of linkages in carbon, nitrogen, and sulfur cycling among widespread estuary sediment bacteria. *Microbiome*, **3**, 14.

Beitel, C. W., Froenicke, L., Lang, J. M., et al. (2014). Strain- and plasmid-level deconvolution of a synthetic metagenome by sequencing proximity ligation products. *PeerJ*, **2**, e415.

Boisvert, S., Raymond, F., Godzaridis, E., Laviolette, F. & Corbeil, J. (2012) Ray Meta: scalable de novo metagenome assembly and profiling. *Genome Biology*, **13**, R122.

Bowe, A., Onodera, T., Sadakane, K. & Shibuya, T. (2012) Succinct de Bruijn graphs. In: B. Raphael & J. Tang (eds), *Algorithms in Bioinformatics*. Springer, Berlin.

Brown, C. T., Hug, L. A., Thomas, B. C., et al. (2015) Unusual biology across a group comprising more than 15% of domain Bacteria. *Nature*, **523**, 208–211.

Buchfink, B., Xie, C. & Huson, D. H. (2015) Fast and sensitive protein alignment using DIAMOND. *Nature Methods*, **12**, 59–60.

Burton, J. N., Liachko, I., Dunham, M. J. & Shendure, J. (2014) Species-level deconvolution of metagenome assemblies with Hi-C-based contact probability maps. *G3: Genes|Genomes|Genetics*, **4**, 1339–1346.

Caspi, R., Billington, R., Ferrer, L., et al. (2016) The MetaCyc database of metabolic pathways and enzymes and the BioCyc collection of pathway/genome databases. *Nucleic Acids Research*, **44**, D471–480.

Cleary, B., Brito, I. L., Huang, K., et al. (2015) Detection of low-abundance bacterial strains in metagenomic datasets by eigengenome partitioning. *Nature Biotechnology*, **33**, 1053–1060.

Crowe, S. A., Maresca, J. A., Jones, C., et al. (2014) Deep-water anoxygenic photosynthesis in a ferruginous chemocline. *Geobiology*, **12**, 322–339.

Crusoe, M. R., Alameldin, H. F., Awad, S., et al. (2015) The khmer software package: enabling efficient nucleotide sequence analysis. *F1000Research*, **4**, 900.

Delong, E. F., Preston, C. M., Mincer, T., et al. (2006) Community genomics among stratified microbial assemblages in the ocean's interior. *Science*, **311**, 496–503.

Denef, V. J. & Banfield, J. F. (2012) In situ evolutionary rate measurements show ecological success of recently emerged bacterial hybrids. *Science*, **336**, 462–466.

Denef, V. J., Kalnejais, L. H., Mueller, R. S., et al. (2010a) Proteogenomic basis for ecological divergence of closely related bacteria in natural acidophilic microbial communities. *Proceedings of the National Academy of Sciences of the United States of America*, **107**, 2383–2390.

Denef, V. J., Mueller, R. S. & Banfield, J. F. (2010b) AMD biofilms: using model communities to study microbial evolution and ecological complexity in nature. *ISME Journal*, **4**, 599–610.

Haider, B., Ahn, T. H., Bushnell, B., Chai, J., Copeland, A. & Pan, C. (2014) Omega: an overlap-graph de novo assembler for metagenomics. *Bioinformatics*, **30**, 2717–2722.

Handley, K. M., Bartels, D., O'Loughlin, E. J., et al. (2014) The complete genome sequence for putative H(2)- and S-oxidizer Candidatus Sulfuricurvum sp., assembled de novo from an aquifer-derived metagenome. *Environmental Microbiology*, **16**, 3443–3462.

Howe, A. & Chain, P. S. (2015) Challenges and opportunities in understanding microbial communities with metagenome assembly (accompanied by IPython Notebook tutorial). *Frontiers in Microbiology*, **6**, 678.

Howe, A. C., Jansson, J. K., Malfatti, S. A., Tringe, S. G., Tiedje, J. M. & Brown, C. T. (2014) Tackling soil diversity with the assembly of large, complex metagenomes.

Proceedings of the National Academy of Sciences of the United States of America, **111**, 4904–4909.

Hug, L. A., Thomas, B. C., Sharon, I., et al. (2016) Critical biogeochemical functions in the subsurface are associated with bacteria from new phyla and little studied lineages. *Environmental Microbiology,* **18**, 159–173.

Hultman, J., Waldrop, M. P., Mackelprang, R., et al. (2015) Multi-omics of permafrost, active layer and thermokarst bog soil microbiomes. *Nature,* **521**, 208–212.

Huson, D. H., Auch, A. F., Qi, J. & Schuster, S. C. (2007) MEGAN analysis of metagenomic data. *Genome Research,* **17**, 377–386.

Iverson, V., Morris, R. M., Frazar, C. D., Berthiaume, C. T., Morales, R. L. & Armbrust, E. V. (2012) Untangling genomes from metagenomes: revealing an uncultured class of marine Euryarchaeota. *Science,* **335**, 587–590.

Kanehisa, M., Sato, Y., Kawashima, M., Furumichi, M. & Tanabe, M. (2016) KEGG as a reference resource for gene and protein annotation. *Nucleic Acids Research,* **44**, D457–462.

Kashtan, N., Roggensack, S. E., Rodrigue, S., et al. (2014) Single-cell genomics reveals hundreds of coexisting subpopulations in wild Prochlorococcus. *Science,* **344**, 416–420.

Langmead, B. & Salzberg, S. L. (2012) Fast gapped-read alignment with Bowtie 2. *Nature Methods,* **9**, 357–359.

Lawrence, J. G. & Ochman, H. (1998) Molecular archaeology of the Escherichia coli genome. *Proceedings of the National Academy of Sciences of the United States of America,* **95**, 9413–9417.

Lesniewski, R. A., Jain, S., Anantharaman, K., Schloss, P. D. & Dick, G. J. (2012) The metatranscriptome of a deep-sea hydrothermal plume is dominated by water column methanotrophs and lithotrophs. *ISME Journal,* **6**, 2257–2268.

Li, D., Liu, C. M., Luo, R., Sadakane, K. & Lam, T. W. (2015) MEGAHIT: an ultra-fast single-node solution for large and complex metagenomics assembly via succinct de Bruijn graph. *Bioinformatics,* **31**, 1674–1676.

Li, H. & Durbin, R. (2009) Fast and accurate short read alignment with Burrows–Wheeler transform. *Bioinformatics,* **25**, 1754–1760.

Lo, I., Denef, V. J., Verberkmoes, N. C., et al. (2007) Strain-resolved community proteomics reveals recombining genomes of acidophilic bacteria. *Nature,* **446**, 537–541.

Luo, C., Tsementzi, D., Kyrpides, N. C. & Konstantinidis, K. T. (2012) Individual genome assembly from complex community short-read metagenomic datasets. *ISME Journal,* **6**, 898–901.

Mavromatis, K., Ivanova, N., Barry, K., et al. (2007) Use of simulated data sets to evaluate the fidelity of metagenomic processing methods. *Nature Methods,* **4**, 495–500.

Meyer, F., Paarmann, D., d'Souza, M., et al. (2008) The metagenomics RAST server – a public resource for the automatic phylogenetic and functional analysis of metagenomes. *BMC Bioinformatics,* **9**, 386.

Muggli, M. D., Bowe, A., Noyes, N. R., et al. (2017) Succinct colored de Bruijn graphs. *Bioinformatics,* **33**, 3181–3187.

Nagarajan, N. & Pop, M. (2013) Sequence assembly demystified. *Nature Reviews Genetics,* **14**, 157–167.

Namiki, T., Hachiya, T., Tanaka, H. & Sakakibara, Y. (2012) MetaVelvet: an extension of Velvet assembler to de novo metagenome assembly from short sequence reads. *Nucleic Acids Research,* **40**, e155.

Orellana, L. H., Rodriguez, R. L. & Konstantinidis, K. T. (2017) ROCker: accurate detection and quantification of target genes in short-read metagenomic data sets by modeling sliding-window bitscores. *Nucleic Acids Research*, **45**, e14.

Overbeek, R., Begley, T., Butler, R. M., et al. (2005) The subsystems approach to genome annotation and its use in the project to annotate 1000 genomes. *Nucleic Acids Research*, **33**, 5691–5702.

Pell, J., Hintze, A., Canino-Koning, R., Howe, A., Tiedje, J. M. & Brown, C. T. (2012) Scaling metagenome sequence assembly with probabilistic de Bruijn graphs. *Proceedings of the National Academy of Sciences of the United States of America*, **109**, 13272–13277.

Peng, Y., Leung, H. C. M., Yiu, S. M. & Chin, F. Y. L. (2010) *IDBA – A Practical Iterative De Bruijn graph De Novo Assembler*. 14th Annual International Conference on Research in Computational Molecular Biology (RECOMB 2010), Lisbon, Portugal. Springer, pp. 426–440.

Peng, Y., Leung, H. C., Yiu, S. M. & Chin, F. Y. (2011) Meta-IDBA: a *de novo* assembler for metagenomic data. *Bioinformatics*, **27**, i94–101.

Peng, Y., Leung, H. C., Yiu, S. M. & Chin, F. Y. (2012) IDBA-UD: a de novo assembler for single-cell and metagenomic sequencing data with highly uneven depth. *Bioinformatics*, **28**, 1420–1428.

Podell, S. & Gaasterland, T. (2007) DarkHorse: a method for genome-wide prediction of horizontal gene transfer. *Genome Biology*, **8**, R16.

Pop, M. (2009) Genome assembly reborn: recent computational challenges. *Briefings in Bioinformatics*, **10**, 354–366.

Reed, D. C., Breier, J. A., Jiang, H., et al. (2015) Predicting the response of the deep-ocean microbiome to geochemical perturbations by hydrothermal vents. *ISME Journal*, **9**, 1857–1869.

Rodriguez-R, L. M. & Konstantinidis, K. T. (2014) Bypassing cultivation to identify bacterial species. *Microbe*, **9**, 111–118.

Rosen, M. J., Davison, M., Bhaya, D. & Fisher, D. S. (2015) Microbial diversity. Fine-scale diversity and extensive recombination in a quasisexual bacterial population occupying a broad niche. *Science*, **348**, 1019–1023.

Rusch, D. B., Halpern, A. L., Sutton, G., et al. (2007) The Sorcerer II Global Ocean Sampling expedition: Northwest Atlantic through eastern tropical Pacific. *PLoS Biology*, **5**, e77.

Satinsky, B. M., Fortunato, C. S., Doherty, M., et al. (2015) Metagenomic and meta-transcriptomic inventories of the lower Amazon River, May 2011. *Microbiome*, **3**, 39.

Scholz, M., Lo, C. C. & Chain, P. S. (2014) Improved assemblies using a source-agnostic pipeline for MetaGenomic Assembly by Merging (MeGAMerge) of contigs. *Science Reports*, **4**, 6480.

Shakya, M., Quince, C., Campbell, J. H., Yang, Z. K., Schadt, C. W. & Podar, M. (2013) Comparative metagenomic and rRNA microbial diversity characterization using archaeal and bacterial synthetic communities. *Environmental Microbiology*, **15**, 1882–1899.

Sharon, I. & Banfield, J. F. (2013) Genomes from metagenomics. *Science*, **342**, 1057–1058.

Sharon, I., Kertesz, M., Hug, L. A., et al. (2015) Accurate, multi-kb reads resolve complex populations and detect rare microorganisms. *Genome Research*, **25**, 534–543.

Sheik, C. S., Jain, S. & Dick, G. J. (2014) Metabolic flexibility of enigmatic SAR324 revealed through metagenomics and metatranscriptomics. *Environmental Microbiology*, **16**, 304–317.

Shi, Y., Tyson, G. W., Eppley, J. M. & Delong, E. F. (2011) Integrated metatranscriptomic and metagenomic analyses of stratified microbial assemblages in the open ocean. *ISME Journal*, **5**, 999–1013.

Simmons, S. L., Dibartolo, G., Denef, V. J., Goltsman, D. S. A., Thelen, M. P. & Banfield, J. F. (2008) Population genomic analysis of strain variation in *Leptospirillum* group II bacteria involved in acid mine drainage formation. *PLoS Biology*, **6**, e177.

Tatusov, R. L., Galperin, M. Y., Natale, D. A. & Koonin, E. V. (2000) The COG database: a tool for genome-scale analysis of protein functions and evolution. *Nucleic Acids Research*, **28**, 33–36.

Thomas, T., Gilbert, J. & Meyer, F. (2012) Metagenomics – a guide from sampling to data analysis. *Microbial Informatics and Experimentation*, **2**, 3.

Tyson, G. W., Chapman, J., Hugenholtz, P., et al. (2004) Community structure and metabolism through reconstruction of microbial genomes from the environment. *Nature*, **428**, 37–43.

Venter, J. C., Remington, K., Heidelberg, J. F., et al. (2004) Environmental genome shotgun sequencing of the Sargasso Sea. *Science*, **304**, 66–74.

Vollmers, J., Wiegand, S. & Kaster, A. K. (2017) Comparing and evaluating metagenome assembly tools from a microbiologist's perspective – not only size matters! *PLoS One*, **12**, e0169662.

Wooley, J. C., Godzik, A. & Friedberg, I. (2010) A primer on metagenomics. *PLoS Computational Biology*, **6**, e1000667.

Wrighton, K. C., Thomas, B. C., Miller, C. S., et al. (2012) Fermentation, hydrogen, and sulfur metabolism in multiple uncultivated bacterial phyla. *Science*, **337**, 1661–1665.

Wu, D., Hugenholtz, P., Mavromatis, K., et al. (2009) A phylogeny-driven genomic encyclopaedia of Bacteria and Archaea. *Nature*, **462**, 1056–1060.

7 Metagenomic Binning

Introduction

Only under the best circumstances will assembly of metagenomic sequences yield complete, closed genomes. This is rare and typically occurs only when high genomic coverage of populations with little strain variation is obtained (Chivian et al. 2008; Schofield et al. 2015)). More frequently, metagenomic assembly produces contigs of short to intermediate length (1–100s of kb). Assigning these contigs to particular microbial groups and/or organisms, a process known as metagenomic binning, offers critical biological and ecological insights. For example, knowing which genes go with which organisms enables an understanding of how different functions are coupled within genomes, thus linking biogeochemical cycles (Walsh et al. 2009); shows how functions are distributed across different organisms in a community, yielding insights into metabolic interactions within communities (Anantharaman et al. 2016a); and provides whole-genome perspectives on microbial evolution (Bendall et al. 2016).

There are two main approaches to binning. The first involves comparing the unknown gene/contig sequences to sequences of known phylogenetic affiliation. This can be done by BLAST or other alignment tools. This approach may work well when the identity of organisms is known and reference genomes closely related to the target populations are available. However, because of the prevalence of horizontal gene transfer and the relatively low number (30–40) of reliable phylogenetic marker genes per genome (Ciccarelli et al. 2006), the vast majority of contigs in a metagenome are difficult to assign by such phylogenetic methods (also see Chapter 8). The second method is to assign contigs to taxonomic groups based on signatures of nucleotide composition such as GC content or tetranucleotide frequency. The overwhelming advantage of using nucleotide composition for binning

Genomic Approaches in Earth and Environmental Sciences, First Edition. Gregory Dick.
© 2019 John Wiley & Sons Ltd. Published 2019 by John Wiley & Sons Ltd.

is that the signal is pervasive throughout the genome, and horizontally transferred genes quickly take on the signal of their host (Lawrence & Ochman 1997). Nucleotide composition-based approaches are particularly valuable for novel genes, which are prevalent in microbial genomes (see Chapter 2) and would not be accurately binned by comparative approaches.

Here we focus primarily on compositional approaches. Note that in a strict sense, "binning," in which sequences are clustered into subgroups that are not necessarily identified taxonomically, should be distinguished from "classification" in which taxonomic labels have been applied. However, as we will see, these processes are often coupled in practice.

7.1   Genomic Signatures of Nucleotide Composition

Remarkably, microorganisms have characteristic patterns of nucleotide composition that are conserved within genomes and distinct between genomes. Such compositional biases were recognized at the very beginning of the age of genomics (Karlin et al. 1997) and metagenomics (Teeling et al. 2004; Tyson et al. 2004). The biases include features such as GC content $((G+C)/(A+T+C+G))$, and the relative abundance of oligonucleotide sequences of a given length (di-, tri-, tetra-, etc. nucleotides), which are illustrated in Figure 7.1. The driving force behind such signatures is still debated but it is related to both neutral (mutation bias due to replication, repair, and recombination) and selective (environmental adaptation,

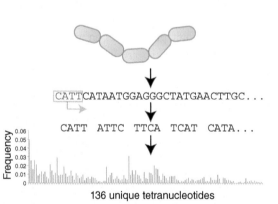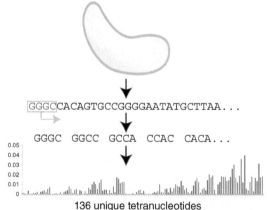

136 unique tetranucleotides   136 unique tetranucleotides

Figure 7.1 Schematic illustration of genome signatures of two different microorganisms. For *tetra*nucleotides, a four-base window is slid one base at a time and the frequency of each four-letter oligonucleotide is counted. The resulting frequency pattern (*histograms at bottom*) is conserved across the whole genome and distinct between genomes, provided sufficient evolutionary divergence. The strand of DNA sampled by metagenomics is disregarded by summing counts from pairs of reverse complementary oligonucleotides together. There are 256 possible tetranucleotides but after summing reverse complements, only 136 are unique.

codon usage, resource availability, energetics) processes (see section 2.3) (Bohlin 2011).

Importantly for metagenomics, strong environmental pressures shared by co-occurring microorganisms do not obscure the signatures between organisms, even in extreme environments (Dick et al. 2009). Also critical is that the signature is pervasive in nearly all genes in a genome. However, recently acquired genes may have the signature of their donor organism (Lawrence & Ochman 1998). Over time, the nucleotide composition signature of horizontally transferred genes is converted to that of the new host during the process of amelioration (Lawrence & Ochman 1997).

In theory, oligonucleotides of any length can be used for binning. What length is optimal? There is a trade-off between information content and length of contig required to provide sufficient data on frequency of specific oligo sequences. Longer oligo patterns contain more information and offer higher specificity (Bohlin et al. 2008). Yet the number of possible oligonucleotides increases exponentially with oligo length: di-, tri-, tetra-, and pentanucleotides have 16, 64, 256, and 1024 possible oligos, respectively (including nonunique reverse complements). Thus, in order to generate a statistically significant frequency histogram, more "samples" are required for longer oligos. In other words, longer contigs are required for longer oligos. Tetranucleotide frequencies typically provide acceptable results down to ~2.5 kb and are very robust above 5 kb, and thus find wide application in current metagenomics projects where contigs in this length range are abundant (Dick et al. 2009). Alternatively, a recently developed approach called ABAWACA allows users to leverage the strengths of multiple lengths of oligonucleotides in a novel iterative algorithm (Brown et al. 2015). A detailed description of this method is not yet published.

7.2 Binning Programs

Binning by nucleotide composition can be approached by either *supervised* or *unsupervised* methods. Supervised methods are trained on the signatures of reference genomes to construct a model that is then used to assign unknown contigs to the reference genomes. This approach is valuable where the organisms of interest are closely related to known reference genomes. Such supervised approaches are often superior in terms of specificity and sensitivity, especially in recognizing signatures in shorter sequence fragments. Their major limitation is the reliance on reference genomes. At this stage of microbial ecology, in which the genomic diversity of many natural microbial communities is still vastly undersampled, this is a major disadvantage. Unsupervised approaches do not rely on *a priori* information about the genome signatures of reference genomes. Rather, they group unknown sequences into clusters directly based on the similarity of their genome signatures. Below, we highlight some of the main strategies used for both supervised and unsupervised approaches through examples of several of the most widely used binning applications.

One of the most widely used supervised methods of metagenomic binning is Phylopithia (McHardy et al. 2007). This method "learns" the characteristics of genome signatures of reference genomes through a support vector machine classifier and then phylogenetically classifies unknown sequences at multiple levels of taxonomy. In addition to nucleotide composition, it also leverages information from phylogenetic marker genes. Phylophythia is highly accurate in terms of both sensitivity and specificity, even for relatively short sequence fragments, though accuracy decreases below 3 kb and especially below 1 kb. The main limitation is that accuracy is lower for clades that are not represented in the "classifier" stage (i.e., reference genomes). A recent update to Phylophythia includes automated methods that enable higher throughput analyses across multiple metagenomes (Gregor et al. 2016). It is also readily integrated with differential coverage methods (Albertsen et al. 2013), discussed below, and the mmgenome toolbox and associated R packages (http://madsalbertsen.github.io/mmgenome/).

Naive Bayesian classifiers have also been used to classify short sequence fragments (down to 400 bp) from genomes of pure cultures at an accuracy of 85% (Sandberg et al. 2001). Another approach uses interpolated Markov models to classify sequences down to 100 bp (Brady & Salzberg 2009). This approach can be combined with alignment (BLAST)-based methods and classifies sequences at all taxonomic levels in all three domains of life (Brady & Salzberg 2011). While quite accurate at higher levels, it is less accurate at finer scales such as the genus level (~70–80%).

Unsupervised methods of binning are indispensable in cases where suitable reference genomes are not available. A key aspect of unsupervised approaches is the visualization of discrete clusters of nucleotide compositional space that signify genomic bins. One particularly effective method of clustering and visualizing sequences is self-organizing maps, which were first pioneered for binning purposes by Abe and colleagues (Abe et al. 2003, 2005). Dick et al. (2009) showed that this approach works well for metagenomes containing novel microbial taxa with no *a priori* genomic information by evaluating the binning process using independently assembled genomes as the "answer key."

The introduction of emergent self-organizing maps (Ultsch & Moerchen 2005) for the purpose of binning improved visualization of genomic bins (Dick et al. 2009). In ESOM, the difference in genome signature between sequence fragments (e.g., Euclidean distance in tetranucleotide frequency) is represented by different colors on a map that serves as a background for the data points. Thus, small differences within genomes are visualized as cohesive genomics bins, and large differences between genomes are visualized as barriers between bins. Though this process is completely unsupervised, including reference genomes as internal standards facilitates interpretation and linking of environmental populations with known organisms. This approach is useful not only for resolving completely novel taxa, but also for recognizing genome elements that are distinct yet related to the core genome signature, such as plasmids, phage, and other parts of the flexible genome (Anantharaman et al. 2014, 2016a; Dick et al. 2009). Recent papers report further improvements in accuracy and visualization of metagenomic binning (Laczny et al. 2015).

A major disadvantage of ESOM (and most current binning approaches) is that it is not fully automated. The capability to recover hundreds of genomes from one metagenome is already routine and recovery of thousands of genomes per metagenome is now being reported (Anantharaman et al. 2016b). Considering that current sequencing throughput now enables the sequencing of dozens of such metagenomes at a time, it becomes apparent that manual analysis of such metagenomic binning data quickly becomes intractable. Automated methods are required to deal with this data deluge, and recently developed applications provide just that.

MaxBin (and MaxBin2) provides an automated pipeline that uses an Expectation-Maximization algorithm, which is reported to push the sequence length threshold down to 1000 bp (Wu et al. 2014, 2016). MetaBat also uses a combination of tetranucleotide frequency and abundance (coverage) information and is reported to be fully automated (Kang et al. 2015). This approach has been used in the binning of metagenomic data from permafrost microbial communities (Hultman et al. 2015). Anvi'o (Eren et al. 2015) uses the automated Concoct method and is highly valuable for its visualization and manual curation capabilities, which allow detailed evaluation of binning evidence, refinement of bins, and comparison of binning results from different methods. Methods such as DAS Tool are critical for the efficient analysis of large complex metagenomic datasets because they enable the comparison and merging of bins from multiple samples and methods (Sieber et al. 2017). A recent review of binning approaches provides further detail and discussion of approaches for metagenomic binning (Sangwan et al. 2016).

A summary of some of the most popular binning programs is shown in Table 7.1. Publication of these new approaches often comes with claims of improvements in accuracy, speed, and visualization of binning. However, in some cases relative performance may depend on the dataset, and in other cases the "performance" can be somewhat subjective. In general, there is a need for more rigorous and comprehensive evaluation and benchmarking of these binning approaches on a variety of different datasets that span a range of diversity, complexity, and taxonomic composition.

7.3 Additional Signal and Steps for Binning: Coverage, Taxonomic Data, and Mini-Assemblies

In addition to nucleotide composition, another valuable factor for binning is the differential abundance of organisms across multiple samples. This idea rests on the fact that (i) all fragments of a given genome should be present at

Table 7.1 Selected methods for metagenomic binning.

Method	Oligo length	Minimum contig length	Type	Notes and references
Phylophythia (S+)	4	400 bp	Supervised	McHardy et al. 2007; Gregor et al. 2016
PhymmBL		100 bp	Supervised	Brady & Salzberg 2009, 2011
Tetra-ESOM	4	2.5 kb	Unsupervised	Effective visualization; not fully automated; Ultsch & Moerchen 2005; Dick et al. 2009
TETRA	4	N.D.	Unsupervised	Measures departure of signature from random expectation and compares via Pearson correlation coefficient; Teeling et al. 2004
MaxBin2	4	1 kb	Unsupervised	Fully automated; also uses differential coverage; Wu et al. 2016
VizBin	4	1 kb	Unsupervised	Laczny et al. 2015
MetaBat	4	1.5 kb	Unsupervised	Kang et al. 2015; Hultman et al. 2015
Concoct	4	1 kb	Unsupervised	Alneberg et al. 2014
ABAWACA	Multiple	1 kb	Unsupervised	Brown et al. 2015
Anvi'o	4	N.A.	Unsupervised/ manual finishing	Outstanding visualization of results, manual refinement, and comparison of results from different methods; Eren et al. 2015

N.A., not applicable; N.D., not determined.

roughly equal abundance in a sample (though see caveats below); (ii) the abundance of organisms (and their genomes) is likely to vary across different samples; (iii) this variance in abundance is likely to be distinct for different organisms. Organism abundance in metagenomic datasets is tracked by mapping reads to contigs to calculate coverage. Indeed, the value of coverage for metagenomic binning was demonstrated in one of the very first metagenomic sequencing efforts (Tyson et al. 2004). Differential coverage can be used in combination with nucleotide composition to enhance the resolution of binning (Alneberg et al. 2014; Sharon & Banfield 2013; Wrighton et al. 2012). Other approaches that fully rely on differential abundance have also been reported to be highly successful (Albertsen et al. 2013; Nielsen et al. 2014). Tutorials and tools that embrace this multi-metagenome concept are now available (http://madsalbertsen.github.io/multi-metagenome/; http://madsalbertsen.github.io/mmgenome/).

Several potential complications should be kept in mind when using differential coverage for binning. First, technological sequencing biases can result in uneven sequence coverage, resulting in artificial coverage variance that could theoretically confound differential coverage binning. For example, some technologies have trouble sequencing extreme nucleotide composition (i.e., GC-rich) and produce lower coverage in those genomic regions. Second, repetitive genomic elements often co-assemble (see Chapters 5 and 6), producing "pile-ups" with artificially high coverage. Third, in actively growing microorganisms the chromosomal copy number is in fact not equal across the whole genome. During DNA replication, which proceeds bidirectionally from a point of origin towards a single terminus, genes already passed by the replication fork will have a higher copy number than as yet unreplicated genes. Thus, variation in coverage *within* genomes of up to three-fold can be expected, especially for fast-growing microbes. This metric has even been proposed as means of estimating growth rate from metagenomic data (Brown et al. 2016; Korem et al. 2015). The variance in coverage due to these three issues is generally low relative to the variance in genomic abundance between samples, but it may limit the resolution gained by using differential coverage information.

Binning can be used as a framework for performing several steps that can facilitate the improvement and analysis of metagenomic assemblies. One increasingly common strategy is to use the contigs associated with a specific genome bin through initial binning to extract all DNA sequence reads from that bin for subsequent reassembly. This can be done by mapping all reads from a metagenomic dataset onto the binned contigs of interest and seems to improve the assemblies substantially (Hug et al. 2013, 2016b). Mate-pair or paired-end information can also be used to evaluate the fidelity of bins and or add contigs to bins (Boetzer et al. 2011; Sekiguchi et al. 2015). This process can be facilitated through visualization strategies such as cytoscape (Shannon et al. 2003). Finally, because binning is often conducted on small fragments that represent a subset of each contig (e.g., a 5 kb window) (Dick et al. 2009), binning can be used to identify assembly errors. A contig that has pieces that fall into two different bins is likely chimeric; binning can quickly flag such contigs for further curation.

7.4 Identifying, Evaluating, and Assessing the Completeness of Genomic Bins

Identification of genomic bins leads to a series of questions regarding their composition and identity. How complete is each bin (i.e., what fraction of the genome was captured)? What is the taxonomic identity of these organisms? Because bins are sometimes resolved at the species to genus level, there may very well be multiple genotypes present, so we also want to know how many different genomes/strains are contained within each bin.

Ideally, identification of genomic bins is done via the 16S rRNA gene, the gold standard of phylogenetic markers. The 16S rRNA gene can be found by annotating the contigs (see Chapter 8) or by searching contigs by BLAST against a database of 16S rRNA genes or with a rRNA predictor such as barrnap (https://github.com/tseemann/barrnap). Because the automated annotators often miss fragmented 16S rRNA genes, the latter is recommended.

However, there are two more fundamental issues that often thwart this effort: (i) because it does not encode protein, the 16S rRNA gene does not hold the genome signature of nucleotide composition and so sequence fragments containing the 16S rRNA gene are often not binned properly; (ii) due to its high sequence conservation, the 16S rRNA gene is often misassembled, including chimeras and consequent termination of contigs, resulting in contigs that are too short to bin. The situation with the 23S rRNA is slightly better in that it is less conserved but it still faces the same issues. Thus, protein-coding phylogenetic marker genes are highly valuable because of their greater sequence heterogeneity and conformity to the genome-wide signature of nucleotide composition that is used for binning. Ribosomal protein S3 (Baker et al. 2015; Hug et al. 2013) and concatenated ribosomal proteins (Hug et al. 2016a) have been used in this capacity.

Only a very small subset of genes in a genome is suitable for conducting universal phylogenetic analysis and evaluating bin completeness. Ideally, genes to be used for such analyses would be (i) universally present in all microorganisms in a single copy, and (ii) not frequently horizontally transferred, so as to reflect the core evolutionary signal of an organism. A set of 31 such genes was originally identified by Ciccarelli et al. (2006) and is frequently used in assessing genome completeness. However, there are downsides of using such a small set of marker genes in that it constitutes a very small portion of the genome, and the genes are typically distributed unevenly in a genome (Sharon & Banfield 2013). One solution is to relax the requirements for defining universal single-copy genes. Rinke et al. (2013) defined a set of 139 bacterial and 162 archaeal conserved single-copy genes on the basis of their presence in a single copy in 90% of genomes as identified by hits to the Pfam database. Another strategy is to define conserved single-copy genes on a per phylum basis (Swan et al. 2013). Lineage-specific markers have also been applied to taxonomic profiling of whole communities (Segata et al. 2012) (see Chapter 8).

Lineage-specific analysis of phylogenetic marker genes was recently integrated into CheckM, an automated method for assessing the completeness and contamination of bacterial and archaeal genomes (Parks et al. 2015). The reference genomes used to generate marker gene sets are determined by phylogenetic analysis. Such lineage-specific marker gene sets offer better estimates of completeness and contamination than universal or domain-specific marker gene sets. Of course, this depends on the availability of phylogenetically related reference genomes; for some novel deeply branching phyla, which are remarkably common (Brown et al. 2015), domain-specific marker sets must be used. CheckM also assesses the number of strains within a bin (a common occurrence in metagenomic bins), and distinguishes this from contamination (i.e., erroneous assignment of contigs to a bin). The automated nature of this program makes it valuable for high-throughput analysis of large datasets. Another automated method, Phylosift, uses a phylogenetic approach and can also be applied to metagenomic bins (Darling et al. 2014).

References

Abe, T., Kanaya, S., Kinouchi, M., Ichiba, Y., Kozuki, T. & Ikemura, T. (2003) Informatics for unveiling hidden genome signatures. *Genome Research*, **13**, 693–702.

Abe, T., Sugawara, H., Kinouchi, M., Kanaya, S. & Ikemura, T. (2005) Novel phylogenetic studies of genomic sequence fragments derived from uncultured microbe mixtures in environmental and clinical samples. *DNA Research*, **12**, 281–290.

Albertsen, M., Hugenholtz, P., Skarshewski, A., Nielsen, K. L., Tyson, G. W. & Nielsen, P. H. (2013) Genome sequences of rare, uncultured bacteria obtained by differential coverage binning of multiple metagenomes. *Nature Biotechnology*, **31**, 533–538.

Alneberg, J., Bjarnason, B. S., de Bruijn, I., et al. (2014) Binning metagenomic contigs by coverage and composition. *Nature Methods*, **11**, 1144–1146.

Anantharaman, K., Duhaime, M. B., Breier, J. A., Wendt, K. A., Toner, B. M. & Dick, G. J. (2014) Sulfur oxidation genes in diverse deep-sea viruses. *Science*, **344**, 757–760.

Anantharaman, K., Breier, J. A. & Dick, G. J. (2016a) Metagenomic resolution of microbial functions in deep-sea hydrothermal plumes across the Eastern Lau Spreading Center. *ISME Journal*, **10**, 225–239.

Anantharaman, K., Brown, C. T., Hug, L. A., et al. (2016b) Thousands of microbial genomes shed light on interconnected biogeochemical processes in an aquifer system. *Nature Communications*, **7**, 13219.

Baker, B. J., Lazar, C., Teske, A. & Dick, G. J. (2015) Genomic resolution of linkages in carbon, nitrogen, and sulfur cycling among widespread estuary sediment bacteria. *Microbiome*, **3**, 14.

Bendall, M. L., Stevens, S. L., Chan, L. K., et al. (2016) Genome-wide selective sweeps and gene-specific sweeps in natural bacterial populations. *ISME Journal*, **10**, 1589–1601.

Boetzer, M., Henkel, C. V., Jansen, H. J., Butler, D. & Pirovano, W. (2011) Scaffolding pre-assembled contigs using SSPACE. *Bioinformatics*, **27**, 578–579.

Bohlin, J. (2011) Genomic signatures in microbes – properties and applications. *Scientific World Journal*, **11**, 715–725.

Bohlin, J., Skjerve, E. & Ussery, D. W. (2008) Investigations of oligonucleotide usage variance within and between prokaryotes. *PLoS Computational Biology*, **4**, e1000057.

Brady, A. & Salzberg, S. L. (2009) Phymm and PhymmBL: metagenomic phylogenetic classification with interpolated Markov models. *Nature Methods*, **6**, 673–676.

Brady, A. & Salzberg, S. (2011) PhymmBL expanded: confidence scores, custom databases, parallelization and more. *Nature Methods*, **8**, 367.

Brown, C. T., Hug, L. A., Thomas, B. C., et al. (2015) Unusual biology across a group comprising more than 15% of domain Bacteria. *Nature*, **523**, 208–211.

Brown, C. T., Olm, M. R., Thomas, B. C. & Banfield, J. F. (2016) Measurement of bacterial replication rates in microbial communities. *Nature Biotechnology*, **34**, 1256–1263.

Chivian, D., Brodie, E. L., Alm, E. J., et al. (2008) Environmental genomics reveals a single-species ecosystem deep within Earth. *Science*, **322**, 275–278.

Ciccarelli, F. D., Doerks, T., von Mering, C., Creevey, C. J., Snel, B. & Bork, P. (2006) Toward automatic reconstruction of a highly resolved tree of life. *Science*, **311**, 1283–1287.

Darling, A. E., Jospin, G., Lowe, E., et al. (2014) PhyloSift: phylogenetic analysis of genomes and metagenomes. *PeerJ*, **2**, e243.

Dick, G. J., Andersson, A. F., Baker, B. J., et al. (2009) Community-wide analysis of microbial genome sequence signatures. *Genome Biology*, **10**, R85.

Eren, A. M., Esen, O. C., Quince, C., et al. (2015) Anvi'o: an advanced analysis and visualization platform for 'omics data. *PeerJ*, **3**, e1319.

Gregor, I., Droge, J., Schirmer, M., Quince, C. & Mchardy, A. C. (2016) PhyloPythiaS+: a self-training method for the rapid reconstruction of low-ranking taxonomic bins from metagenomes. *PeerJ*, **4**, e1603.

Hug, L. A., Castelle, C. J., Wrighton, K. C., et al. (2013) Community genomic analyses constrain the distribution of metabolic traits across the Chloroflexi phylum and indicate roles in sediment carbon cycling. *Microbiome*, **1**, 22.

Hug, L. A., Baker, B. J., Anantharaman, K., et al. (2016a) A new view of the tree of life. *Nature Microbiology*, **1**, 16048.

Hug, L. A., Thomas, B. C., Sharon, I., et al. (2016b) Critical biogeochemical functions in the subsurface are associated with bacteria from new phyla and little studied lineages. *Environmental Microbiology*, **18**, 159–173.

Hultman, J., Waldrop, M. P., Mackelprang, R., et al. (2015) Multi-omics of permafrost, active layer and thermokarst bog soil microbiomes. *Nature*, **521**, 208–212.

Kang, D. D., Froula, J., Egan, R. & Wang, Z. (2015) MetaBAT, an efficient tool for accurately reconstructing single genomes from complex microbial communities. *PeerJ*, **3**, e1165.

Karlin, S., Mrázek, J. & Campbell, A. M. (1997) Compositional biases of bacterial genomes and evolutionary implications. *Journal of Bacteriology*, **179**, 3899–3913.

Korem, T., Zeevi, D., Suez, J., et al. (2015) Growth dynamics of gut microbiota in health and disease inferred from single metagenomic samples. *Science*, **349**, 1101–1106.

Laczny, C. C., Sternal, T., Plugaru, V., et al. (2015) VizBin – an application for reference-independent visualization and human-augmented binning of metagenomic data. *Microbiome*, **3**, 1.

Lawrence, J. G. & Ochman, H. (1997) Amelioration of bacterial genomes: rates of change and exchange. *Journal of Molecular Evolution*, **44**, 383–397.

Lawrence, J. G. & Ochman, H. (1998) Molecular archaeology of the Escherichia coli genome. *Proceedings of the National Academy of Sciences of the United States of America*, **95**, 9413–9417.

McHardy, A. C., Martín, H. G., Tsirigos, A., Hugenholtz, P. & Rigooutsos, I. (2007) Accurate phylogenetic classification of variable-length DNA fragments. *Nature Methods*, **4**, 63–72.

Nielsen, H. B., Almeida, M., Juncker, A. S., et al. (2014) Identification and assembly of genomes and genetic elements in complex metagenomic samples without using reference genomes. *Nature Biotechnology*, **32**, 822–828.

Parks, D. H., Imelfort, M., Skennerton, C. T., Hugenholtz, P. & Tyson, G. W. (2015) CheckM: assessing the quality of microbial genomes recovered from isolates, single cells, and metagenomes. *Genome Research*, **25**, 1043–1055.

Rinke, C., Schwientek, P., Sczyrba, A., et al. (2013) Insights into the phylogeny and coding potential of microbial dark matter. *Nature*, **499**, 431–437.

Sandberg, R., Winberg, G., Bränden, C., Kaske, A., Ernberg, I. & Cöster, J. (2001) Capturing whole-genome characteristics in short sequences using a naive bayesian classifier. *Genome Research*, **11**, 1404–1409.

Sangwan, N., Xia, F. & Gilbert, J. A. (2016) Recovering complete and draft population genomes from metagenome datasets. *Microbiome*, **4**, 8.

Schofield, M. M., Jain, S., Porat, D., Dick, G. J. & Sherman, D. H. (2015) Identification and analysis of the bacterial endosymbiont specialized for production of the chemotherapeutic natural product ET-743. *Environmental Microbiology*, **17**, 3964–3975.

Segata, N., Waldron, L., Ballarini, A., Narasimhan, V., Jousson, O. & Huttenhower, C. (2012) Metagenomic microbial community profiling using unique clade-specific marker genes. *Nature Methods*, **9**, 811–814.

Sekiguchi, Y., A., O., Parks, D. H., Yamauchi, T., Tyson, G. W. & Hugenholtz, P. (2015) First genomic insights into members of a candidate bacterial phylum responsible for wastewater bulking. *PeerJ*, **3**, e740.

Shannon, P., Markiel, A., Ozier, O., et al. (2003) Cytoscape: a software environment for integrated models of biomolecular interaction networks. *Genome Research*, **13**, 2498–2504.

Sharon, I. & Banfield, J. F. (2013) Genomes from metagenomics. *Science*, **342**, 1057–1058.

Sieber, C. M. K., Probst, A. J., Sharrar, A. M., et al. (2017) Recovery of genomes from metagenomes via a dereplication, aggregation, and scoring strategy. *bioRxiv*. Available at: www.biorxiv.org/content/early/2017/02/11/107789 (accessed 31 October 2017).

Swan, B. K., Tupper, B., Sczyrba, A., et al. (2013) Prevalent genome streamlining and latitudinal divergence of planktonic bacteria in the surface ocean. *Proceedings of the National Academy of Sciences of the United States of America*, **110**, 11463–11468.

Teeling, H., Meyerdierks, A., Bauer, M., Amann, R. & Glöckner, F. O. (2004) Application of tetranucleotide frequencies for the assignment of genomic fragments. *Environmental Microbiology*, **6**, 938–947.

Tyson, G. W., Chapman, J., Hugenholtz, P., et al. (2004) Community structure and metabolism through reconstruction of microbial genomes from the environment. *Nature*, **428**, 37–43.

Ultsch, A. & Moerchen, F. (2005) *ESOM-Maps: Tools for Clustering, Visualization, and Classification with Emergent SOM*. Technical Report. Department of Mathematics and Computer Science, University of Marburg, Germany, p. 46.

Walsh, D. A., Zaikova, E., Howes, C. G., et al. (2009) Metagenome of a versatile chemolithoautotroph from expanding oceanic dead zones. *Science*, **326**, 578–582.

Wrighton, K. C., Thomas, B. C., Miller, C. S., et al. (2012) Fermentation, hydrogen, and sulfur metabolism in multiple uncultivated bacterial phyla. *Science*, **337**, 1661–1665.

Wu, Y. W., Tang, Y. H., Tringe, S. G., Simmons, B. A. & Singer, S. W. (2014) MaxBin: an automated binning method to recover individual genomes from metagenomes using an expectation-maximization algorithm. *Microbiome*, **2**, 26.

Wu, Y. W., Simmons, B. A. & Singer, S. W. (2016) MaxBin 2.0: an automated binning algorithm to recover genomes from multiple metagenomic datasets. *Bioinformatics*, **32**, 605–607.

8

Annotation

Gene Calling, Taxonomy, and Function

Introduction

Metagenomic datasets are valuable for determining the taxonomic and/or functional composition of microbial communities. Several approaches can be taken to achieve these goals, and the choice of which approach to use depends on the scientific question as well as the nature of the dataset.

As discussed earlier, a key question is whether to utilize assembled contigs and/or genomic bins or to work at the individual DNA sequence read level (see sections 4.1 and 6.1). Even in cases where assembly and genomic binning are conducted, a substantial portion of the dataset is often unassembled and/or unbinned, and is thus left over as individual reads or as contigs that are too short for binning (Fig. 8.1). The proportion of the dataset that ends up in each of these pools varies from small to overwhelming majority, depending on the amount and type of sequencing conducted, the taxonomic and genomic diversity of the community being sequenced, and the tools employed for analyses. In cases where many reads are unassembled or in contigs that are too short to bin, analysis at the read level may be appropriate. Where bins are available, there are tremendous advantages to utilizing the information they provide (see section 6.1). Hence, this chapter covers both assembly/bin-dependent and bin-independent methods for the tasks of taxonomic and functional studies of metagenomes.

Genomic Approaches in Earth and Environmental Sciences, First Edition. Gregory Dick.
© 2019 John Wiley & Sons Ltd. Published 2019 by John Wiley & Sons Ltd.

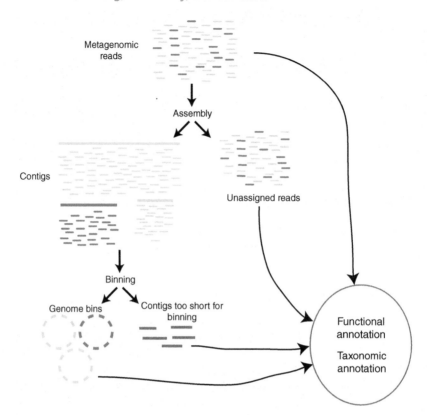

Figure 8.1 Schematic illustration of workflow for a metagenomic sequencing project with emphasis on materials used for taxonomic and functional annotation. Short thin lines represent individual sequencing reads, with different colours from different populations, and thicker longer lines represent consensus sequences of contigs. Assembly of sequencing reads into contigs and genomic bins is preferable, but even where that is possible, a large portion of the dataset is often on short contigs or individual reads that do not assemble. Thus, functional and taxonomic annotation of individual reads is also desirable.

8.1 Gene Calling

After genome assembly and binning, genes are identified within the contigs. For protein-coding genes, identification of gene start and stop sites is a key first step towards predicting amino acid sequences and inferring function and taxonomy. A nice overview of this process is provided in Thomas et al. (2012). After identification of genes by start and stop site, functions are assigned (see sections 8.1 and 8.2). Gene coordinates (start and stop sites) for protein-coding genes are generally determined by inherent features such as start and stop codons and ribosome binding sites, whereas RNAs are determined by comparison to databases, often with hidden markov models.

A challenge with applying methods developed for clonal genomes to metagenomes is that they may rely on reference genomes for training and/or codon usage information, whereas metagenomes contain multiple genomes from multiple species. In addition, short sequences such as individual sequence reads or short contigs present challenges associated with sequencing errors and partial genes (Trimble et al. 2012). The Trimble et al. study showed that the best choice of gene prediction tool depends on the nature of the data: FragGeneScan for raw reads (Kim et al. 2015; Rho et al. 2010) and Prodigal (Hyatt et al. 2010, 2012) for assembled contigs. Genemark (Borodovsky et al. 2003), metagenemark (Zhu et al. 2010), and Prokka (Seemann 2014) are also competitive. In addition to using the intrinsic characteristics for the *ab initio* process described above, the new NCBI Prokaryotic Genome Automatic Annotation Pipeline method now integrates extrinsic information in the form of comparative analysis (e.g., alignment-based protein predictions) (Tatusova et al. 2016).

Although the accuracy of gene-calling processes is generally quite high, a significant portion of genes in unassembled metagenomic datasets can still be missed (Thomas et al. 2012). Assembly of reads into contigs facilitates gene calling. Pipelines such as MG-RAST (Meyer et al. 2008), IMG/M-ER (Markowitz et al. 2009), and Metapathways (Hanson et al. 2014) bring together numerous processes for calling of various types of genes and features (e.g., CRISPRS, rRNAs, tRNAs, protein-coding genes) and synthesize results from search of multiple functional databases (see section 8.3).

8.2 Determining Taxonomic Composition

Determining which microorganisms are present in a sample is routinely done by sequencing a gene that is present in all target organisms and of sufficiently conserved sequence and length that it can be reliably amplified and compared between organisms and to databases of sequences of known taxonomy for classification (e.g., PCR and sequencing of the 16S rRNA gene; see section 1.2). Although 16S rRNA gene sequence remains an effective and efficient method, there are several reasons why taxonomic analysis of shotgun sequence datasets is also valuable. First, so-called "universal" PCR primers for the 16S rRNA gene may introduce biases or even miss novel organisms (Baker et al. 2006; Brown et al. 2015). Second, whole-genome approaches can offer taxonomic data at finer scales, since sequences of numerous genes with higher resolving power than the 16S are produced. Third, in some projects the shotgun data is being produced anyway (e.g., for functional purposes); this may circumvent the need to do PCR-based studies. Finally, during the course of omics projects, it is often desirable to identify the taxonomy of particular genomic fragments (e.g., a functional gene of special interest) or bins that do not carry a 16S sequence.

The task of taxonomic identification can come in many flavors, involving a variety of different goals and types of data. For example, we may want to identify a particular genomic bin. Or, at the other end of the spectrum, we may want to characterize an entire community by taxonomically profiling shotgun sequence data at the read level. Not surprisingly, such diverse questions are addressed with a diverse array of approaches. Below we highlight some of the challenges in conducting taxonomic identification on shotgun data and then outline some of the approaches and their strengths and weaknesses.

Several important challenges are encountered when performing taxonomic identification using shotgun sequence data.

* Unlike sequences produced by PCR, shotgun sequences have arbitrary start and stop sites, complicating their alignment.
* Almost all genes display much more sequence heterogeneity than the 16S gene, so the vast majority of genes in shotgun sequence data will be poorly conserved and potentially difficult to compare between distantly related organisms. This is important because microbes within natural microbial communities are often quite novel, so the most closely related sequences in databases may offer little resolving power for classification.
* There are few publicly available databases that offer highly curated taxonomic information, such as those available for 16S rRNA sequences (SILVA – Pruesse et al. (2007), RDP – Cole et al. (2009), and Greengenes – DeSantis et al. (2006)).

A common yet crude method of taxonomic assignment is taxonomic/phylogenetic profiling, in which all sequences (reads or called genes/predicted proteins) within a dataset are compared (e.g., with BLAST) to a database containing sequences of known taxonomic affiliation, and the taxonomic distribution of best matches is tabulated. This approach is implemented as a tool in some of the commonly used publicly available analysis pipelines such as MG-RAST (Meyer et al. 2008) and Integrated Microbial Genomes (Markowitz et al. 2009), and is valuable as a "quick and dirty" view, but it has substantial shortcomings. First, only a very small portion of the dataset will be captured. Because of the extensive genomic diversity discussed earlier (see Chapter 2), there are typically many genes in natural communities that are not present in databases. Second, there is the "best hit problem" in which a sequence hits many different taxonomic groups with near-equal similarity. It would be assigned to the "best hit" but the difference between matches might not be statistically significant. This best hit issue can be resolved by employing a lowest common ancestor algorithm as described in the program MEGAN (Huson et al. 2007). However, this approach often leads to low-resolution assignments (i.e., at very high taxonomic levels) due to the inherent limitations of the databases. Another issue is that different proteins evolve at different rates, and thus the degree of sequence conservation varies enormously between and even within protein families. Thus, applying the same alignment threshold to very different proteins is not ideal.

Aside from these issues that are inherent to the sequences and databases themselves, there are also limitations in the algorithms used to compare sequences to each other. For example, BLAST sequence similarity scores do not necessarily reflect phylogenetic relationships (Smith & Pease 2017). BLAST is also increasingly unfeasible due to the large computational requirements (and long compute times) required to analyze the large metagenomic datasets that are now routinely produced. Methods that avoid the computationally expensive step of inexact sequence alignment are much faster. For example, Kraken uses exact matches of short sequences (k-mers) (Wood & Salzberg 2014), and MASH efficiently estimates distances between pair-wise sequences (Ondov et al. 2016).

The whole-genome taxonomic profiling methods described above are coarse because they assign equal weight to the taxonomic signal of each gene in a genome. However, the degree to which a gene accurately reflects an organism's evolutionary history varies widely; some genes have evolved together largely in one organism, in an evolutionary core, whereas others may be frequently horizontally transferred between organisms, obscuring their vertical evolutionary history. Genes that offer the most phylogenetic information are those that are (i) present in all (or most) organisms, (ii) rarely horizontally transferred, (iii) of sufficient length. Such genes are often a part of critical cellular processes such as translation, which is conducted by all cellular life and involves highly specific functions of interacting molecules in a way that the protein sequences must be highly conserved. As discussed in Chapter 7, these genes are invaluable in identifying metagenomics bins. Phylogenetic marker genes are also increasingly used to assess the taxonomic composition of microbial communities, through analysis of sequences at either the read or contig level.

As described above, a challenge of analyzing phylogenetic marker genes derived from shotgun sequence data is that the sequences have arbitrary start and stop sites. Thus, they may be only partially overlapping or even nonoverlapping, complicating efforts to construct sequence alignments, for example. A solution is to map onto reference phylogenies, a task for which effective methods have been developed (Matsen et al. 2010; Segata et al. 2012; von Mering et al. 2007). One of these methods, pplacer, was recently integrated into PhyloSift, a pipeline for phylogenetic analysis of metagenomic data (Darling et al. 2014).

Another challenge of analyzing taxonomic composition is that reference phylogenies (or genomes) vastly underrepresent the diversity present in natural microbial communities (Baker & Dick 2013). In fact, a recent study showed that reference-based approaches can miss about half of species abundance and richness (Sunagawa et al. 2013). Thus, reference-independent methods for resolving the diversity inherent in the environment are required. An early solution was provided in the context of functional analysis by grouping protein-coding sequences into operational protein families, analogous to operational taxonomic units (OTUs) (Schloss & Handelsman 2008). More recently, Sunagawa et al. (2013) developed a method for forming metagenomic operational taxonomic units (mOTUs) with single-copy phylogenetic marker genes.

8.3 Functional Annotation

The goal of many studies that apply omics approaches to questions in the Earth and environmental sciences is to uncover the roles of microorganisms in biogeochemical processes. Achievement of this goal requires that we assign physiological and biochemical functions to sequences, a process referred to as functional annotation. This can be done at either the level of individual sequence reads or after assembly of these reads into contigs (see Fig. 8.1). Annotation of contigs is preferable; it is generally more sensitive and accurate because full-length genes are available, but some high-diversity communities remain refractory to assembly, necessitating read-based approaches.

Functional annotation is a major challenge and only 20–50% of genes in a metagenomic dataset are typically assigned a function (Gilbert et al. 2010). Indeed, there is a huge amount of sequence data that has not been assigned functions (Yooseph et al. 2007); a substantial fraction of the proteins these genes encode are essential (Goodacre et al. 2014), and this gap in knowledge is a major challenge to the field (Godzik 2011). Because links between sequence and function are almost always derived from genetic and biochemical studies of cultured organisms, these so-called "culture-independent" methods are absolutely dependent on existing knowledge from cultured organisms! Hence, there is appeal in using clustering-based approaches that do not rely on functional annotations, but allow ecological analysis to be conducted on the basis of homology (Schloss & Handelsman 2008). However, such approaches have not yet seen the same widespread use as the functional annotation methods discussed below.

8.3.1 Overall Approach to Functional Annotation

In contrast to the mainly *ab initio* approach used for gene calling, functional annotations are typically done through comparison of the sequences to databases of gene or protein sequences with functional annotations. Comparisons are generally with one of two methods: BLAST (or some derivative, like BLAT) (Kent 2002) and hidden markov models (Eddy 2004). Since no reference database covers all biological functions, and each database has strengths and weaknesses, many popular pipelines such as MG-RAST (Meyer et al. 2008), IMG/M-ER (Markowitz et al. 2009), and InterProScan (Jones et al. 2014) draw on searches against multiple databases and then merge them into a single framework. These databases include KEGG (Kanehisa et al. 2016), eggNOG (Muller et al. 2010), COG/KOG (Tatusov et al. 2003), Pfam (Finn et al. 2016), and TIGRFAM (Haft et al.

2013). Thomas et al. (2012) and Teeling and Glockner (2012) discuss these methods in more detail. Functional annotations from one or more of these databases or pipelines can then be used in combination with a variety of statistical methods to compare the functions of microbial genomes or communities (see Chapter 12). Recently developed software packages also synthesize the annotations from isolate genomes or genomic bins to predict phenotype, including traits such as carbon and energy sources, sporulation, antibiotic susceptibility, and enzymatic activities (Weimann et al. 2016).

An entirely different strategy to investigate the function of microbial communities is to make predictions based on phylogenetic marker genes such as 16S rRNA, combined with knowledge of the functional capacity of closely related organisms. This approach has been pioneered by Curtis Huttenhower and colleagues with the software PICRUSt (Langille et al. 2013). It relies on correlation between function and phylogeny, which, as discussed in the section on genomic diversity (see Chapter 2), is not always conserved. However, the power of this method is that the confidence (or uncertainty) of predictions can be estimated by evaluating the link between phylogeny and function in reference genomes. While such an approach is no replacement for tracking functions with omics sequencing, it does provide potential value for making predictions and developing hypotheses, perhaps as one tier in a cost-effective process where screening a large number of samples with marker genes is then followed with shotgun sequencing of a subset of samples.

8.3.2   *Predicting Metabolic Pathways*

Annotation of functions to individual genes often gets us only part way to the goal of uncovering microbial physiology, metabolisms, and roles in biogeochemical cycling. The enzymes encoded by genes catalyze biochemical reactions that typically operate in concert with other enzymes within metabolic pathways. Because some proteins can belong to multiple pathways while others are unique (and thus more informative) (Caspi et al. 2013; Hanson et al. 2014), putting functional annotations into the context of the metabolic pathways in which they participate can be profitable. Of course, some pathways occur within a single organism, but there are also pathways that span multiple organisms via interspecies metabolic interactions. These interactions can be integral to biogeochemical processes, so investigating pathways in the context of whole microbial communities offers one means of identifying such interactions (Hanson et al. 2014).

Several methods are commonly used for analyzing metabolic pathways of microbial genomes and communities. Prediction of metabolic pathways remains challenging, especially determining directionality and specificity, so the output of these methods should be interpreted with caution (Caspi et al. 2013). The KEGG pathway is a collection of manually drawn metabolic maps that represent knowledge of pathways in model organisms (Kanehisa et al. 2014). Genomic information from the nonmodel organisms being studied is then projected onto these maps in an attempt to associate their genes with the experimentally derived links between sequence and function underlying the metabolic maps. This projection is done through KEGG Orthology (KO) groups using methods of defining orthology described above. Because the KEGG pathway is based on a limited number of model organisms, it does not have flexibility to accommodate variation in metabolic pathways, and tends to have somewhat coarse resolution (Altman et al. 2013). The KEGG Automatic Annotation Server (KAAS) is an online service that uses BLAST to compare sequences against the manually curated KEGG GENES database, producing KO assignments and automatically generated KEGG pathways (www.genome.jp/tools/kaas/). KEGG is also implemented in numerous metagenomic pipelines and downstream applications for comparing the functional composition of microbial communities (Abubucker et al. 2012).

MetaCyc is another database of metabolic pathways based on experimental information (Caspi et al. 2016). Compared to KEGG, it uses smaller, more evolutionarily conserved units to represent the sequence–function link. MetaCyc interfaces with BioCyc, a collection of databases representing the genome and predicted metabolic networks for more than 3000 organisms (ranging from highly curated to automated). These databases are associated with the software package PathwayTools, which enables users to predict metabolic networks of a sequenced genome (Karp et al. 2011). Together, these tools provide an integrated structure for investigating metabolic pathways, complete with descriptions of the pathways and associated enzymes and literature citations so that users can critically assess predictions. While this extensive documentation and the finer resolution of MetaCyc offer considerable advantages over KEGG, in the experience of the author's lab the predicted pathways are often relatively sparse, suggesting a trade-off between sensitivity and specificity. However, this lack of sensitivity likely more accurately reflects the lack of experimental data and knowledge regarding the function of many genes within microbial communities.

An exciting development is the application of MetaCyc to whole environmental sequence data through the MetaPathways annotations and analysis pipeline (Hanson et al. 2014). The MetaPathways pipeline offers (i) prediction of protein-coding open reading frames (ORFs) via Prodigal, (ii) annotation using the MetaCyc, RefSeq, KEGG, and COG protein databases, (iii) taxonomic analysis using MEGAN, ML-TreeMap, 16S SSU and 23S LSU rRNA homology using the Silva and GreenGenes databases, and (iv) systematic creation of Environmental Pathway/Genome Databases (ePGDBs) mapping functional information onto the MetaCyc database of metabolic pathways.

8.3.3 The Importance of Experimental Annotation

Ultimately, the quality of functional annotation depends on the knowledge linking genes and proteins to biochemical and physiological functions. Because this knowledge is woefully inadequate, there is an urgent need for continued and renewed experimental efforts aimed at identifying gene functions. These efforts include conventional genetic and biochemical approaches as well as high-throughput methods that can be applied on the whole-genome scale (see section 3.3).

References

Abubucker, S., Segata, N., Goll, J., et al. (2012) Metabolic reconstruction for metagenomic data and its application to the human microbiome. *PLoS Computational Biology*, **8**, e1002358.

Altman, T., Travers, M., Kothari, A., Caspi, R. & Karp, P. D. (2013) A systematic comparison of the MetaCyc and KEGG pathway databases. *BMC Bioinformatics*, **14**, 112.

Baker, B. J. & Dick, G. J. (2013) Omic approaches in microbial ecology: charting the unknown. *Microbe*, **8**, 353–360.

Baker, B. J., Tyson, G. W., Webb, R. I., et al. (2006) Lineages of acidophilic archaea revealed by community genomic analysis. *Science*, **314**, 1933–1935.

Borodovsky, M., Mills, R., Besemer, J. & Lomsadze, A. (2003) Prokaryotic gene prediction using GeneMark and GeneMark.hmm. *Current Protocols in Bioinformatics*, Chapter 4, Unit 4 5.

Brown, C. T., Hug, L. A., Thomas, B. C., et al. (2015) Unusual biology across a group comprising more than 15% of domain Bacteria. *Nature*, **523**, 208–211.

Caspi, R., Dreher, K. & Karp, P. D. (2013) The challenge of constructing, classifying, and representing metabolic pathways. *FEMS Microbiology Letters*, **345**, 85–93.

Caspi, R., Billington, R., Ferrer, L., et al. (2016) The MetaCyc database of metabolic pathways and enzymes and the BioCyc collection of pathway/genome databases. *Nucleic Acids Research*, **44**, D471–480.

Cole, J. R., Wang, Q., Cardenas, E., et al. (2009) The Ribosomal Database Project: improved alignments and new tools for rRNA analysis. *Nucleic Acids Research*, **37**, D141–145.

Darling, A. E., Jospin, G., Lowe, E., et al. (2014) PhyloSift: phylogenetic analysis of genomes and metagenomes. *PeerJ*, **2**, e243.

DeSantis, T. Z., Hugenholtz, P., Larsen, N., et al. (2006) Greengenes, a chimera-checked 16S rRNA gene database and workbench compatible with ARB. *Applied and Environmental Microbiology*, **72**, 5069–5072.

Eddy, S. R. (2004) What is a hidden Markov model? *Nature Biotechnology*, **22**, 1315–1316.

Finn, R. D., Coggill, P., Eberhardt, R. Y., et al. (2016) The Pfam protein families database: towards a more sustainable future. *Nucleic Acids Research*, **44**, D279–285.

Gilbert, J. A., Field, D., Swift, P., et al. (2010) The taxonomic and functional diversity of microbes at a temperate coastal site: a 'multi-omic' study of seasonal and diel temporal variation. *PLoS One*, **5**, e15545.

Godzik, A. (2011) Metagenomics and the protein universe. *Current Opinion in Structural Biology*, **21**, 398–403.

Goodacre, N. F., Gerloff, D. L. & Uetz, P. (2014) Protein domains of unknown function are essential in bacteria. *mBio*, **5**, e00744–13.

Haft, D. H., Selengut, J. D., Richter, R. A., Harkins, D., Basu, M. K. & Beck, E. (2013) TIGRFAMs and genome properties in 2013. *Nucleic Acids Research*, **41**, D387–395.

Hanson, N. W., Konwar, K. M., Hawley, A. K., Altman, T., Karp, P. D. & Hallam, S. J. (2014) Metabolic pathways for the whole community. *BMC Genomics*, **15**, 619.

Huson, D. H., Auch, A. F., Qi, J. & Schuster, S. C. (2007) MEGAN analysis of metagenomic data. *Genome Research*, **17**, 377–386.

Hyatt, D., Locascio, P. F., Hauser, L. J. & Uberbacher, E. C. (2012) Gene and translation initiation site prediction in metagenomic sequences. *Bioinformatics*, **28**, 2223–2230.

Hyatt, D., Chen, G. L., Locascio, P. F., Land, M. L., Larimer, F. W. & Hauser, L. J. (2010) Prodigal: prokaryotic gene recognition and translation initiation site identification. *BMC Bioinformatics*, **11**, 119.

Jones, P., Binns, D., Chang, H. Y., et al. (2014) InterProScan 5: genome-scale protein function classification. *Bioinformatics*, **30**, 1236–1240.

Kanehisa, M., Goto, S., Sato, Y., Kawashima, M., Furumichi, M. & Tanabe, M. (2014) Data, information, knowledge and principle: back to metabolism in KEGG. *Nucleic Acids Research*, **42**, D199–205.

Kanehisa, M., Sato, Y., Kawashima, M., Furumichi, M. & Tanabe, M. (2016) KEGG as a reference resource for gene and protein annotation. *Nucleic Acids Research*, **44**, D457–462.

Karp, P. D., Latendresse, M. & Caspi, R. (2011) The pathway tools pathway prediction algorithm. *Standards in Genomic Science*, **5**, 424–429.

Kent, W. J. (2002) BLAT – the BLAST-like alignment tool. *Genome Research*, **12**, 656–664.

Kim, D., Hahn, A. S., Wu, S., Hanson, N. W., Konwar, K. M. & Hallam, S. J. (2015) *FragGeneScan-plus for scalable high-throughput short-read open reading frame prediction.* Computational Intelligence in Bioinformatics and Computational Biology, IEEE Conference, Niagara Falls, ON, pp. 1–8.

Langille, M. G., Zaneveld, J., Caporaso, J. G., et al. (2013) Predictive functional profiling of microbial communities using 16S rRNA marker gene sequences. *Nature Biotechnology*, **31**, 814–821.

Markowitz, V. M., Mavromatis, K., Ivanova, N. N., Chen, I. M., Chu, K. & Kyrpides, N. C. (2009) IMG ER: a system for microbial genome annotation expert review and curation. *Bioinformatics*, **25**, 2271–2278.

Matsen, F. A., Kodner, R. B. & Armbrust, E. V. (2010) pplacer: linear time maximum-likelihood and Bayesian phylogenetic placement of sequences onto a fixed reference tree. *BMC Bioinformatics*, **11**, 538.

Meyer, F., Paarmann, D., d'Souza, M., et al. (2008) The metagenomics RAST server – a public resource for the automatic phylogenetic and functional analysis of metagenomes. *BMC Bioinformatics*, **9**, 386.

Muller, J., Szklarczyk, D., Julien, P., et al. (2010) eggNOG v2.0: extending the evolutionary genealogy of genes with enhanced non-supervised orthologous groups, species and functional annotations. *Nucleic Acids Research*, **38**, D190–195.

Ondov, B. D., Treangen, T. J., Melsted, P., et al. (2016) MASH: fast genome and metagenome distance estimation using MinHash. *Genome Biology*, **17**, 132.

Pruesse, E., Quast, C., Knittel, K., et al. (2007) SILVA: a comprehensive online resource for quality checked and aligned ribosomal RNA sequence data compatible with ARB. *Nucleic Acids Research*, **35**, 7188–7196.

Rho, M., Tang, H. & Ye, Y. (2010) FragGeneScan: predicting genes in short and error-prone reads. *Nucleic Acids Research*, **38**, e191.

Schloss, P. D. & Handelsman, J. (2008) A statistical toolbox for metagenomics: assessing functional diversity in microbial communities. *BMC Bioinformatics*, **9**, 34.

Seemann, T. (2014) Prokka: rapid prokaryotic genome annotation. *Bioinformatics*, **30**, 2068–2069.

Segata, N., Waldron, L., Ballarini, A., Narasimhan, V., Jousson, O. & Huttenhower, C. (2012) Metagenomic microbial community profiling using unique clade-specific marker genes. *Nature Methods*, **9**, 811–814.

Smith, S. A. & Pease, J. B. (2017) Heterogenous molecular processes among the causes of how sequence similarity scores can fail to recapitulate phylogeny. *Briefings in Bioinformatics*, **18**, 451–457.

Sunagawa, S., Mende, D. R., Zeller, G., et al. (2013) Metagenomic species profiling using universal. *Nature Methods*, **10**, 1196–1199.

Tatusov, R. L., Fedorova, N. D., Jackson, J. D., et al. (2003) The COG database: an updated version includes eukaryotes. *BMC Bioinformatics*, **4**, 41.

Tatusova, T., Dicuccio, M., Badretdin, A., et al. (2016) NCBI prokaryotic genome annotation pipeline. *Nucleic Acids Research*, **44**, 6614–6624.

Teeling, H. & Glockner, F. O. (2012) Current opportunities and challenges in microbial metagenome analysis – a bioinformatic perspective. *Briefings in Bioinformatics*, **13**, 728–742.

Thomas, T., Gilbert, J. & Meyer, F. (2012) Metagenomics – a guide from sampling to data analysis. *Microbial Informatics and Experimentation*, **2**, 3.

Trimble, W. L., Keegan, K. P., d'Souza, M., et al. (2012) Short-read reading-frame predictors are not created equal: sequence error causes loss of signal. *BMC Bioinformatics*, **13**, 183.

Von Mering, C., Hugenholtz, P., Raes, J., et al. (2007) Quantitative phylogenetic assessment of microbial communities in diverse environments. *Science*, **315**, 1126–1130.

Weimann, A., Mooren, K., Frank, J., Pope, P. B., Bremges, A. & McHardy, A. C. (2016) From genomes to phenotypes: Traitar, the microbial trait analyzer. *mSystems*, **1**, e00101–16.

Wood, D. E. & Salzberg, S. L. (2014) Kraken: ultrafast metagenomic sequence classification using exact alignments. *Genome Biology*, **15**, R46.

Yooseph, S., Sutton, G., Rusch, D. B., et al. (2007) The *Sorcerer II* Global Ocean Sampling expedition: expanding the universe of protein families. *PLoS Biology*, **5**, e16.

Zhu, W., Lomsadze, A. & Borodovsky, M. (2010) Ab initio gene identification in metagenomic sequences. *Nucleic Acids Research*, **38**, e132.

9 Metatranscriptomics

Introduction

Messenger RNA (mRNA) is a short-lived intermediate between DNA and protein. Because transcription of mRNA is often tightly regulated, and the lifetime of mRNA molecules (~minutes) is short with respect to the timescale of environmental changes, the characterization of mRNAs is a powerful method for investigating the cellular response to environmental conditions experienced by microorganisms.

(Meta)transcriptomic approaches seek to probe the transcriptional activity of microbes by characterizing the pool of RNAs that compose the transcriptome of a culture, population, or community. This is challenging for a number of reasons. Samples must be collected quickly and in a way that prevents artificial changes to the transcriptome during sampling. The cellular inventory of mRNA is low – much lower than that of rRNAs, genes, or proteins (Moran et al. 2013). It should also be recognized that there is a poor correspondence between the abundance of transcripts and their corresponding proteins. This is chiefly due to the difference in lifetimes of mRNAs and proteins; most proteins persist long after the mRNAs have been degraded. However, additional factors such as posttranscriptional processing and regulation and variable translation efficiencies may also contribute to the mismatch between mRNA and protein abundance. Hence, transcriptomics and proteomics (see Chapter 10) provide two different types of information; transcriptomics largely reflects *current* cellular demands or environmental signals, whereas proteomics more accurately reflects the standing stock of cellular proteins and their associated catalytic potential, which integrates protein synthesis over a longer timescale. Both types of information are valuable for tracking microbial activity. With current technologies, transcriptomics is substantially more sensitive, generating many

Genomic Approaches in Earth and Environmental Sciences, First Edition. Gregory Dick.
© 2019 John Wiley & Sons Ltd. Published 2019 by John Wiley & Sons Ltd.

more identifiable sequence reads than MS/MS spectra per unit cost, thus providing more information especially for genes that are expressed at lower levels.

One of the outcomes of plunging costs of DNA sequencing has been that so-called RNAseq approaches, in which a pool of cDNA is randomly shotgun sequenced, have overtaken microarrays as the go-to method for conducting transcriptomics, including studies of gene expression in microbial communities. For microbiome studies, this has the tremendous advantage that *all* transcripts are sequenced, so the diversity inherent to natural microbial communities is captured (see Chapter 2). This is in contrast to microarray technologies that require probes of *known* genes. Hence, RNAseq approaches provide a window into community gene expression that is both sensitive to novel genes and holds a wealth of information regarding the variation of gene sequence. In the overview of transcriptomics below, we consider critical issues regarding sampling collection and library preparation, normalization, and downstream applications. Statistical analyses that are common to other forms of data are considered in Chapter 12.

9.1 Sample Collection

Because of the short lifetime of mRNAs, care must be taken to avoid changes in the metatranscriptome during sampling. Concerns include (i) transcriptional shifts that could occur due to changing conditions experienced during sampling, e.g., in response to changes in light availability between the natural sample and a bottle; (ii) degradation of mRNA; and (iii) differential degradation of RNA between different community members (Stewart 2013). A detailed discussion of challenges and methods of sampling and preservation of community RNA from marine planktonic samples is available in Stewart (2013). Many of the considerations discussed therein are applicable to samples from any environment.

Regardless of sample type, minimizing the time between sample collection and preservation is critical. Samples can be preserved by snap freezing in liquid nitrogen or immersion in a reagent such as RNAlater (Stewart 2013). Snap freezing has been found to preserve RNA with comparable effectiveness to other preservatives (Bachoon et al. 2001). However, when working in remote or extreme environments, RNAlater is convenient because it does not need to be frozen immediately and preserves DNA as well as RNA. For example, Breier et al. (2014) developed a sampler for deep-sea hydrothermal vent plumes that filters water and immediately doses the filter with RNAlater *in situ*. *In situ* preservation is also critical to capture variation that occurs on short timescales. The Environmental Science Processor is an autonomous instrument (Herfort et al. 2016) that can also use RNAlater as a preservative of RNA for metatranscriptomics (Ottesen et al. 2011). A risk of using preservatives such as RNAlater, however, is that they may introduce biases in the observed results of sequencing-based analyses (McCarthy et al. 2015; Reck et al. 2015).

9.2 RNA Extraction and Preparation of cDNA Libraries

RNA must be converted to cDNA prior to sequencing, and the methods used for this process can have substantial impacts on the results of metatranscriptomics studies (Alberti et al. 2014; Kratz & Carninci 2014). There are now a number of commercially available kits for preparation of cDNA sequencing libraries, some of which can take as little as 1 ng of RNA as starting material. However, less starting material means more amplification, which usually comes with a higher risk of artifacts. A high degree of technical variation has been observed between metatranscriptomics replicates, underscoring the need for replication and statistics in the assessment of biological variation of gene expression in the environment (Tsementzi et al. 2014). Stewart (2013) and Sarode et al. (2016) provide in-depth looks at methods for RNA extraction and cDNA synthesis for metatranscriptomics.

9.2.1 Should rRNAs Be Removed Prior to Library Preparation and Sequencing?

Only 5–20% of cellular RNA is mRNA, with the balance being dominated by rRNAs. Thus, since mRNA is usually the target, methods for the removal of rRNAs have been developed in order to save sequencing costs. Typically, this is done by some form of subtractive hybridization. Although this was a cost-effective procedure in the early days of metatranscriptomic sequencing, careful consideration of the full costs of labor and supplies shows that it can be more cost-efficient not to remove rRNAs (Stewart 2013). This approach has the added benefit that rRNAs can be used as a measure of relative metabolic activity of OTUs (Lesniewski et al. 2012; Urich et al. 2008), and any artifacts of sample handling during the rRNA removal process are avoided. Hence, the decision on whether to remove rRNAs prior to sequencing depends on the goals of the study and costs that are particular to specific facilities (e.g., labor). The author's lab currently uses the RiboZero rRNA removal kits (Illumina).

9.3 Assigning Transcripts to Genes or Other Features

As is the case with metagenomic reads, short metatranscriptomic reads are difficult to annotate and classify directly. An effective approach is to map reads to longer reference genes. These reference genes can be from reference genomes (Rivers et al. 2013), paired metagenomes (Anantharaman et al. 2013; Lesniewski et al. 2012), or databases of genes, either publicly available (e.g., RefSeq) (Rivers et al. 2013) or custom databases of functional genes of interest (Satinsky et al. 2014) (Fig. 9.1). In cases where the reference genes are closely related to the transcripts (as in the case of paired metagenomes and metatranscriptomes), read mapping programs such as BowTie (Langmead & Salzberg 2012) and the Burrows–Wheeler Aligner (BWA) can be used. The tremendous advantage of this and similar methods is that they are *fast* relative to BLAST. The disadvantages are that (i) they are optimized for close matches (e.g., 2–3 mismatches over a 100 bp read), (ii) BWA is not "competitive," meaning that a read will be assigned to the first database entry found to satisfy the search criteria, not necessarily the best match. BowTie2 does use a quantitative scoring system and it performs well in terms of speed and accuracy relative to other mapping programs (Langmead & Salzberg 2012). Thus, where differentiating expression of closely related genes is desired, cDNA read mapping is best done with a tool such as BLAST or BowTie2.

Where relationships between transcripts and genes are more distant, such as results when comparing environmental transcripts to the NCBI RefSeq database, more sensitive search programs such as BLAST must be used. Here, a major challenge is determining the appropriate cut-off (i.e., bitscore, alignment length, and/or % identity) for "assigning transcripts to genes," a choice complicated by the fact that the relationship between sequence divergence and function is different for each gene (see Chapter 8). New faster algorithms such as BLAT (Kent 2002), LAST (Kielbasa et al. 2011), and RAPSearch2 (Zhao et al. 2012) may prove useful for these purposes. Once transcripts have been assigned to genes, these genes can be incorporated into broader levels of functional classification, such as metabolic pathways, using databases such as KEGG, etc. (Rivers et al. 2013).

9.4 *De Novo* Assembly

A potentially major disadvantage of any database-dependent approach is that transcripts for which there is no corresponding database entry will not be classified (see Fig. 9.1). This issue is particularly acute when using public databases that do not accurately represent the target microbial community. While use of paired metagenomes for analysis of metatranscriptomic datasets helps minimize this problem, it does not completely eliminate it because metagenomes never capture the full gene content of microbial communities. The same risks exist for analysis of metagenomic data, but the stakes are even higher for metatranscriptomic data, where the possibility exists for highly expressed transcripts from low-abundance organisms.

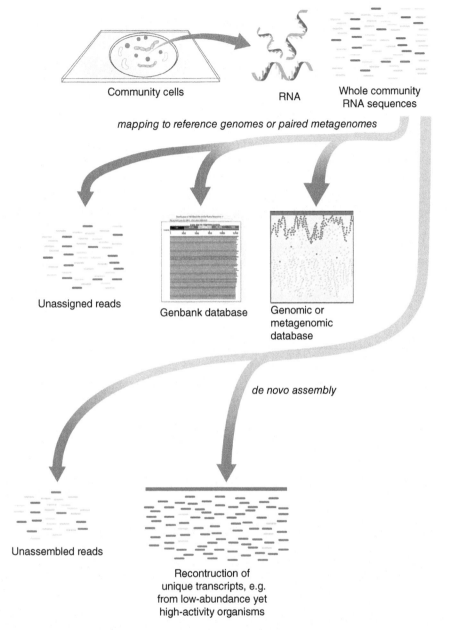

Community cells

RNA

Whole community
RNA sequences

mapping to reference genomes or paired metagenomes

Unassigned reads

Genbank database

Genomic or
metagenomic
database

de novo assembly

Unassembled reads

Recontruction of
unique transcripts, e.g.
from low-abundance yet
high-activity organisms

Figure 9.1 Annotation of transcripts in metatranscriptomics sequencing projects. In order to match a transcript with its corresponding gene, and to facilitate annotation, transcripts are often mapped to genes from reference genomes or from paired metagenomes. The disadvantage of this approach is that genes and associated contigs and genomes may be missing from the metagenome. Another option is to perform *de novo* genome assembly on the metatranscriptomic reads.

One potential solution to this issue is direct *de novo* assembly of metatranscriptomic data (Baker et al. 2013; Celaj et al. 2014; Peng et al. 2013). Each mRNA transcribed from an operon contains several co-transcribed genes, and each operon-derived transcript is physically separate and does not often overlap substantially with other transcripts, so cDNA reads derived from these transcripts will typically only assemble into short fragments of maximum length equal to operon size. However, even these relatively short cDNA contigs represent units of analysis that are a significant improvement in data format and reduction in data complexity over raw reads. Assembly of these operon-level transcriptional units, and subsequent quantification of reads belonging to them, enables the identification of highly transcribed genes and operons that are not present in genomic databases. Such genes can then be manually curated or studied with respect to variance in abundance across environmental or experimental parameters for functional insights. Indeed, Baker et al. (2013) used *de novo* metatranscriptomic assembly to find that some critical biogeochemical functions, such as nitrite oxidation, are mediated by organisms and genes that are at low abundance yet highly expressed.

9.5 Absolute Versus Relative Abundance and Normalization

As discussed above, shotgun sequencing (as opposed to microarrays) is now the dominant approach taken to conduct metatranscriptomics. Like other applications of DNA sequencing to microbial ecology (clone libraries, amplicon sequencing, shotgun metagenomics), metatranscriptomics produces data in terms of *relative* abundance, i.e., the number of reads observed for each gene (or any genetic unit) is dependent on sequencing effort and expressed as a portion of the total number of sequences generated. This complicates comparisons of gene expression between samples because the number of cDNA sequence reads observed for any one gene in a metatranscriptomic shotgun sequencing dataset is influenced by the relative abundance of transcripts from all other genes in all other organisms of the sampled community (see section 4.2.4 and further discussion below).

To circumvent this issue, *absolute* abundance of transcripts in the sampled community (e.g., number of transcripts per volume of sample) can be estimated by adding "internal mRNA standards" of known quantity to the mRNA pool to be sequenced. Absolute abundances of transcripts for a given gene are then calculated by determining the extent to which the internal standards are diluted by natural mRNAs in the sequence libraries (a dilution factor) and multiplying the observed cDNA reads (also known as "counts") for the gene(s) of interest by the dilution factor (Moran et al. 2013; Satinsky et al. 2013), See Risso et al. (2014) on reliability of commercially available standards.

In cases where internal standards are not used, careful thought must be given to the analysis and interpretation of the data. In particular, how should the data be *normalized*? Before addressing this question, we should first consider the various factors that will influence the number of cDNA reads obtained for a given gene in a metatranscriptomic sequencing dataset. First, obviously, observed counts will depend on the expression level of the gene in a cell (i.e., number of transcripts per gene copy per cell). This gene expression level is often what we seek to investigate. Second, it depends on the abundance of cells expressing the gene. Here, we encounter a much different situation to that of transcriptomic studies of pure cultures, where with only one organism present the observed transcriptome is an aggregation of the transcriptomes of a collection of cells in a culture that have identical (or nearly identical) genome sequences. In the case of cultures, there may be differences in the transcriptome of individual cells, and the observed transcriptome integrates this variation into an average of the whole culture, regardless of the number of cells in that culture. In contrast, in a microbial community the gene(s) of interest may be present in only a small fraction of the cells sampled, and the relative abundance of those cells directly influences the observed abundance of transcripts in a metatranscriptomic dataset. Third, not only does the observed abundance of transcripts for a given gene in a shotgun sequencing dataset depend on transcription levels of that gene and the abundance of that cell, it also depends on transcription level of every other gene and the abundance of every other cell in the community. *Relative* abundance is relative to the abundance of other transcripts in the community, which depends on the same factors discussed above. Finally, the observed counts for a gene depend on its length; when reads are shorter than transcripts, the number of reads recovered is expected to be proportional to gene length.

The above considerations emphasize that normalization is one of the most critical steps of analyzing metatranscriptomic data. This process is where we take into account factors such as gene length, sequencing effort, and potential effects of the composition of the RNA pool, so that transcript data can be compared across sequencing libraries and samples. In a pure culture in which copy number per gene does not vary among cells, a common and intuitive procedure is to calculate the proportion of each gene's cDNA reads relative to the total number of cDNA reads and gene length. This proportion, which can then be compared across samples, is often represented as "reads per kilobase per million reads mapped" (RPKM) (Box 9.1). For paired-end data, "reads" is replaced with the "fragment" representing the paired-ends (FPKM).

Though widely used, RPKM is not universally accepted as an optimal method for interpreting transcriptomics or metatranscriptomic data. For transcriptomic data, normalizing by gene length has been shown to introduce bias (Dillies et al. 2013; Oshlack & Wakefield 2009), and for metatranscriptomic data there is additionally the issue that we sometimes do not know gene lengths. Further, as introduced above, metatranscriptomics data

Box 9.1 Methods for normalizing (meta)transcriptomic data.

Normalization is a critical step in analysis and interpretation of metatranscriptomic datasets when the data is in terms of relative abundance (i.e., internal standards are not used). Here we summarize and discuss some of the most widely used methods for normalization. See the main text and Table 9.1 for further details, discussion, and references.

Normalization for read length

All else being equal, the number of reads retrieved for a gene from shotgun sequencing data will be proportional to the length of the gene. Thus, normalizing for gene length removes this effect and allows comparison of read counts between genes of different lengths.

Normalization for sequencing effort

Intuition suggests that read counts should be adjusted for the amount of sequencing done; if twice as much sequencing is done on one sample versus another, the observed counts should be divided by two for each gene. A simple normalization by number of cDNA reads in the library (and gene length) integrates the effects of both cell abundance and transcription levels, and can be used as a measure of the relative contribution of a gene to a metatranscriptome.

RPKM/FPKM (reads/fragments per kilobase per million reads)

This metric is calculated as follows:

$$RPKM = \frac{(cDNA\,reads\,observed\,per\,gene)}{(gene\,length)(total\,number\,of\,cDNA\,reads\,in\,library)}(10^6)$$

RPKM/FPKM normalizes by read length and sequencing effort. It is widely used for studies of single organisms.*

cDNA/DNA ratio

Calculating the ratio of cDNA reads to DNA reads for each gene has been used to measure the ratio of transcripts to gene copies, thus estimating gene expression levels.*

Normalize by total metatranscriptomic reads mapped to a genomic bin

This method provides a measure of the contribution of the transcripts of each gene relative to the other genes in a genome.*

DESeq

Normalizes the data with a scaling factor computed as the median of the ratio of the read count for each gene divided by its geometric mean across all samples/conditions.

* These methods do not take into account organism abundance and/or the effects of other genes. Thus, the observation of counts for any gene is sensitive to the expression levels of all other genes in the sample. When applied to microbial communities, these methods do not provide an accurate measure of expression (transcripts per gene copy) but rather of the relative abundance of transcripts for each gene relative to all other genes.

is sensitive to issues of proportionality that must be considered (i.e., observed "counts" for each gene are relative and thus sensitive to changes in the composition of the RNA pool) (Robinson & Oshlack 2010). Methods such as RPKM do not account for such issues of relative abundance and are thus not robust for making comparisons of metatranscriptomic data between samples (for a brief example see section 4.2.4). For a full description and evaluation of seven commonly used methods of normalization of RNAseq data (for single organisms), see Dillies et al. (2013). In many cases the difference between these methods lies in whether "total number of reads" refers to reads generated, mapped, or some statistical treatment thereof.

For metatranscriptomic datasets for which cDNA has been obtained from microbial communities containing assemblages of numerous species, data analysis is more complex and depends on the scientific question and goal (Table 9.1). Researchers may be interested in the relative abundance of transcripts as a measure of the relative transcriptional activity of various populations, taking into account the combined effects of both cell abundance and expression. In these cases, normalizing for library size and gene length and calculating the ratio of counts for genes under different conditions can provide valuable biological insights (Lesniewski et al. 2012;

Table 9.1 Goals and strategies for normalizing metatranscriptomic data.

Goal	Measure/methods (see Box 9.1)	Notes	References
Quantify overall transcriptional activity as proportion of all genes in an organism/ community	Relative abundance of transcripts in community; normalize by sequencing effort and gene length; includes RPKM/FPKM	Integrates effects of cell abundance and expression level*	Lesniewski et al. 2012; Sanders et al. 2013; Satinsky et al. 2013
Quantify the relative transcriptional importance of gene(s) within a particular organism	Normalize counts by number of cDNA reads mapped to genomic bin	Isolates each genomic bin; removes effects of genes from other bins	Ottesen et al. 2014
Assess gene expression level; identify differentially expressed genes	cDNA/DNA ratio	For shotgun data, both transcriptomic and genomic data are relative*	Frias-Lopez et al. 2008
	DESeq2	Assumption that most genes are not differentially expressed not well studied for communities	Love et al. 2014
	Normalize to a housekeeping gene	Major assumption of constitutive/constant expression level of housekeeping gene is difficult to verify	Harke et al. 2012; Kleiner 2017

*Indicates that in these methods, the results observed for any one gene are sensitive to transcription level of other genes.

Sanders et al. 2013). In cases where expression (i.e., number of transcripts for gene copy) is the metric of interest, calculating the ratio of cDNA/DNA reads for each gene may be appropriate (Frias-Lopez et al. 2008). However, in calculating cDNA/DNA ratios for shotgun datasets, both the numerator and denominator would be in terms of relative abundance, and thus sensitive to the abundance and expression levels of other genes and organisms in the community.

Another possibility is to normalize counts of cDNA reads for a gene to the counts retrieved for some housekeeping gene that is expected to be constitutively expressed at a constant level. This is commonly done for quantitative PCR (Dheda et al. 2004; Harke et al. 2012) and has been done for proteomics (Ferguson et al. 2005) and metagenomics (Manor and Borenstein 2015; Tsementzi et al. 2016). The concern with applying this method to metatranscriptomic data (i.e., to uncultured organisms) is that it is often difficult or impossible to verify that the requirement for constitutive and constant per-cell expression levels is satisfied (Kleiner 2017).

Where possible, normalization of metatranscriptomic data by number of transcripts mapped to a genomic bin can be valuable (Ottesen et al. 2014). This essentially isolates each bin and provides insights into the transcriptional level of a gene relative to other genes in the genome. While it may also be possible to use reference genomes for the purpose of defining bins (Gifford et al. 2013), this approach may potentially miss genes that are specific to "wild" populations, as discussed above. Satinsky et al. (2014) employ all three of these scaling approaches, drawing on transcript inventories, gene expression ratios, as well as transcripts mapping per taxon for various interpretations.

9.6 Detecting Differential Expression

Ultimately, the goal of many transcriptomic studies is to determine which genes are significantly differently expressed between samples and/or conditions. For all methods of testing for differential expression, replication is critical (see section 4.2.1). One particularly valuable method for assessing differential expression is DESeq (DE for "differential expression"), which is available as an R/Bioconductor package (Anders & Huber 2010; Love et al. 2014). This method makes the assumption that most genes are not differentially expressed and therefore should have similar read counts across all samples. Thus, it normalizes the data with a scaling factor computed as the median of the ratio of the read count for each gene divided by its geometric mean across all samples. This method is now being used for transcriptomics studies and has now been used in a number of metatranscriptomic studies (Rivers et al. 2013; Tsementzi et al. 2014).

There are potential complications of applying DESeq2 to microbial communities, and these appear to have not been explicitly considered in the literature. A critical question regarding the DESeq method of normalization described above is whether the variability in community composition complicates the statistical treatment, in particular whether it violates the assumption that most genes will be expressed at similar levels across all samples. In other words, due to substantial variation in the abundance of organisms (and therefore their genes) across samples, it may not be appropriate to normalize based on the mean of raw read counts. Marchetti et al. (2012) argued that normalization methods such as the trimmed mean of M values (TMM) (a predecessor of DESeq) can help minimize the impacts of varying species abundances on inferring gene expression. To the author's knowledge, this issue has not yet been directly evaluated in the published literature. A recent study on the use of DESeq2 for 16S rRNA gene amplicon datasets found that it can have a high false discovery rate under certain data characteristics (Weiss et al. 2017). Although this method should be used with caution when sequence libraries of vastly different size are being compared, it is the best currently available tool for determining differential abundance in RNAseq data when absolute abundance information is not available.

References

Alberti, A., Belser, C., Engelen, S., et al. (2014) Comparison of library preparation methods reveals their impact on interpretation of metatranscriptomic data. *BMC Genomics*, **15**, 912.

Anantharaman, K., Breier, J. A., Sheik, C. S. & Dick, G. J. (2013) Evidence for hydrogen oxidation and metabolic plasticity in widespread deep-sea sulfur-oxidizing bacteria. *Proceedings of the National Academy of Sciences of the United States of America*, **110**, 330–335.

Anders, S. & Huber, W. (2010) Differential expression analysis for sequence count data. *Genome Biology*, **11**, R106.

Bachoon, D. S., Chen, F. & Hodson, R. E. (2001) RNA recovery and detection of mRNA by RT-PCR from preserved prokaryotic samples. *FEMS Microbiology Letters*, **201**, 127–132.

Baker, B. J., Sheik, C. S., Taylor, C. A., et al. (2013) Community transcriptomic assembly reveals microbes that contribute to deep-sea carbon and nitrogen cycling. *ISME Journal*, **7**, 1962–1973.

Breier, J. A., Gomez-Ibanez, D. A., Sayre-McCord, R. T., et al. (2014) A large volume particulate and water multi-sampler with in situ preservation for microbial and biogeochemical studies. *Deep-Sea Research I*, **94**, 195–206.

Celaj, A., Markle, J., Danska, J. & Parkinson, J. (2014) Comparison of assembly algorithms for improving rate of metatranscriptomic functional annotation. *Microbiome*, **2**, 39.

Dheda, K., Huggett, J. F., Bustin, S. A., Johnson, M. A., Rook, G. & Zumla, A. (2004) Validation of housekeeping genes for normalizing RNA expression in real-time PCR. *Biotechniques*, **37**, 112–114, 116, 118–119.

Dillies, M. A., Rau, A., Aubert, J., et al. (2013) A comprehensive evaluation of normalization methods for Illumina high-throughput RNA sequencing data analysis. *Briefings in Bioinformatics*, **14**, 671–683.

Ferguson, R. E., Carroll, H. P., Harris, A., Maher, E. R., Selby, P. J. & Banks, R. E. (2005) Housekeeping proteins: a preliminary study illustrating some limitations as useful references in protein expression studies. *Proteomics*, **5**, 566–571.

Frias-Lopez, J., Shi, Y., Tyson, G. W., et al. (2008) Microbial community gene expression in ocean surface waters. *Proceedings of the National Academy of Sciences of the United States of America*, **105**, 3805–3810.

Gifford, S. M., Sharma, S., Booth, M. & Moran, M. A. (2013) Expression patterns reveal niche diversification in a marine microbial assemblage. *ISME Journal*, **7**, 281–298.

Harke, M. J., Berry, D. L., Ammerman, J. W. & Gobler, C. J. (2012) Molecular response of the bloom-forming cyanobacterium, Microcystis aeruginosa, to phosphorus limitation. *Microbial Ecology*, **63**, 188–198.

Herfort, L., Seaton, C., Wilkin, M., et al. (2016) Use of continuous, real-time observations and model simulations to achieve autonomous, adaptive sampling of microbial processes with a robotic sampler. *Limnology and Oceanography: Methods*, **14**, 50–67.

Kent, W. J. (2002) BLAT – the BLAST-like alignment tool. *Genome Research*, **12**, 656–664.

Kielbasa, S. M., Wan, R., Sato, K., Horton, P. & Frith, M. C. (2011) Adaptive seeds tame genomic sequence comparison. *Genome Research*, **21**, 487–493.

Kleiner, M. (2017) Normalization of metatranscriptomic and metaproteomic data for differential gene expression analyses: the importance of accounting for organism abundance. Available at: https://peerj.com/preprints/2846/ (accessed 31 October 2017).

Kratz, A. & Carninci, P. (2014) The devil in the details of RNA-seq. *Nature Biotechnology*, **32**, 882–884.

Langmead, B. & Salzberg, S. L. (2012) Fast gapped-read alignment with Bowtie 2. *Nature Methods*, **9**, 357–359.

Lesniewski, R. A., Jain, S., Anantharaman, K., Schloss, P. D. & Dick, G. J. (2012) The metatranscriptome of a deep-sea hydrothermal plume is dominated by water column methanotrophs and lithotrophs. *ISME Journal*, **6**, 2257–2268.

Love, M. I., Huber, W. & Anders, S. (2014) Moderated estimation of fold change and dispersion for RNA-seq data with DESeq2. *Genome Biology*, **15**, 550.

Manor, O. & Borenstein, E. (2015) MUSiCC: a marker genes based framework for metagenomic normalization and accurate profiling of gene abundances in the microbiome. *Genome Biology*, **16**, 53.

Marchetti, A., Schruth, D. M., Durkin, C. A., et al. (2012) Comparative metatranscriptomics identifies molecular bases for the physiological responses of phytoplankton to varying iron availability. *Proceedings of the National Academy of Sciences of the United States of America*, **109**, E317–E325.

McCarthy, A., Chiang, E., Schmidt, M. L. & Denef, V. J. (2015) RNA preservation agents and nucleic acid extraction method bias perceived bacterial community composition. *PLoS One*, **10**, e0121659.

Moran, M. A., Satinsky, B., Gifford, S. M., et al. (2013) Sizing up metatranscriptomics. *ISME Journal*, **7**, 237–243.

Oshlack, A. & Wakefield, M. J. (2009) Transcript length bias in RNA-seq data confounds systems biology. *Biology Direct*, **4**, 14.

Ottesen, E. A., Marin, R. 3rd, Preston, C. M., et al. (2011) Metatranscriptomic analysis of autonomously collected and preserved marine bacterioplankton. *ISME Journal*, **5**, 1881–1895.

Ottesen, E. A., Young, C. R., Gifford, S. M., et al. (2014) Multispecies diel transcriptional oscillations in open ocean heterotrophic bacterial assemblages. *Science*, **345**, 207–212.

Peng, Y., Leung, H. C., Yiu, S. M., Lv, M. J., Zhu, X. G. & Chin, F. Y. (2013) IDBA-tran: a more robust de novo de Bruijn graph assembler for transcriptomes with uneven expression levels. *Bioinformatics*, **29**, i326–334.

Reck, M., Tomasch, J., Deng, Z., et al. (2015) Stool metatranscriptomics: a technical guideline for mRNA stabilisation and isolation. *BMC Genomics*, **16**, 494.

Risso, D., Ngai, J., Speed, T. P. & Dudoit, S. (2014) Normalization of RNA-seq data using factor analysis of control genes or samples. *Nature Biotechnology*, **32**, 896–902.

Rivers, A. R., Sharma, S., Tringe, S. G., Martin, J., Joye, S. B. & Moran, M. A. (2013) Transcriptional response of bathypelagic marine bacterioplankton to the Deepwater Horizon oil spill. *ISME Journal*, **7**, 2315–2329.

Robinson, M. D. & Oshlack, A. (2010) A scaling normalization method for differential expression analysis of RNA-seq data. *Genome Biology*, **11**, R25.

Sanders, J. G., Beinart, R. A., Stewart, F. J., Delong, E. F. & Girguis, P. R. (2013) Metatranscriptomics reveal differences in in situ energy and nitrogen metabolism among hydrothermal vent snail symbionts. *ISME Journal*, **7**, 1556–1567.

Sarode, N., Parris, D. J., Ganesh, S., Seston, S. L. & Stewart, F. J. (2016) Generation and analysis of microbial metatranscriptomes. In: M. V. Yates, C. H. Natkatsu, R. V. Miller & S. D. Pillai (eds), *Manual of Environmental Microbiology*, 4th edn. ASM Press, Washington DC.

Satinsky, B. M., Gifford, S. M., Crump, B. C. & Moran, M. A. (2013) Use of internal standards for quantitative metatranscriptome and metagenome analysis. *Microbial Metagenomics, Metatranscriptomics, and Metaproteomics*, **531**, 237–250.

Satinsky, B. M., Crump, B. C., Smith, C. B., et al. (2014) Microspatial gene expression patterns in the Amazon River Plume. *Proceedings of the National Academy of Sciences of the United States of America*, **111**, 11085–11090.

Stewart, F. J. (2013) Preparation of microbial community cDNA for metatranscriptomic analysis in marine plankton. *Methods in Enzymology*, **531**, 187–218.

Tsementzi, D., Poretsky, R., Rodriguez, R. L., Luo, C. & Konstantinidis, K. T. (2014) Evaluation of metatranscriptomic protocols and application to the study of freshwater microbial communities. *Environmental Microbiology Reports*, **6**, 640–655.

Tsementzi, D., Wu, J., Deutsch, S., et al. (2016) SAR11 bacteria linked to ocean anoxia and nitrogen loss. *Nature*, **536**, 179–183.

Urich, T., Lanzen, A., Qi, J., Huson, D. H., Schleper, C. & Schuster, S. C. (2008) Simultaneous assessment of soil microbial community structure and function through analysis of the meta-transcriptome. *PLoS ONE*, **3**, e2527.

Weiss, S., Xu, Z. Z., Peddada, S., et al. (2017) Normalization and microbial differential abundance strategies depend upon data characteristics. *Microbiome*, **5**, 27.

Zhao, Y., Tang, H. & Ye, Y. (2012) RAPSearch2: a fast and memory-efficient protein similarity search tool for next-generation sequencing data. *Bioinformatics*, **28**, 125–126.

10 Metaproteomics

Introduction

Proteomics offers the ability to inventory the proteins in a microbial culture or community. Unlike genomic and transcriptomic approaches, where *de novo* sequencing is relatively straightforward (at least to produce the raw sequence!), the methods for sequencing proteins are more complicated and more dependent on databases. Nevertheless, the pay-off is worthwhile; because proteins directly mediate biogeochemical processes, quantifying their abundance in the environment is critical to tracking the activity and metabolic status of microorganisms. By integrating this information with assembled genomes (e.g., from metagenomes), we can piece together the picture of who is producing which proteins and catalyzing what biogeochemical functions. This integrated approach – aptly termed "proteogenomics" – was first used in cultures (Jaffe et al. 2004), but was soon successfully applied to communities (Lo et al. 2007; Ram et al. 2005; Wilmes & Bond 2004). Proteogenomics opens up new opportunities to study the physiology, ecology, and evolution of microbial populations and communities (Verberkmoes et al. 2009). Verberkmoes et al. distinguish such proteogenomics, in which identified proteins are placed into the genomic context of individual species, from metaproteomics, in which the identified proteins are not assigned to particular species.

Community proteogenomics was pioneered in acid mine drainage ecosystems (Denef et al. 2010b; Ram et al. 2005), where relatively low-diversity biofilm communities are often dominated by just a few species, allowing reconstruction of essentially complete genomes, even before next-generation sequencing (Tyson et al. 2004). These early forays yielded several far-reaching insights:

Genomic Approaches in Earth and Environmental Sciences, First Edition. Gregory Dick.
© 2019 John Wiley & Sons Ltd. Published 2019 by John Wiley & Sons Ltd.

- 50% of the proteome of the dominant organism could be identified
- many proteins of unknown function, so-called "hypothetical proteins," were highly expressed, including some of the most abundant proteins in the entire community, showing that these proteins are real and likely ecologically relevant
- biochemical fractionation of cell components allowed proteins to be localized to cellular fractions (i.e., membrane, soluble, extracellular)
- mutations and posttranslational modifications could be inferred by patterns of peptide mapping
- abundant yet unknown proteins could be targeted for purification and characterization (Ram et al. 2005; Singer. et al. 2008). Recent reviews provide an overview of early community and meta-/community-proteomics studies (Abraham et al. 2014; Verberkmoes et al. 2009).

Proteomics is now being applied to complex microbial communities and across large scales (Hawley et al. 2014; Morris et al. 2010). Saito et al. (2014) recently investigated nutrient stress-related proteins across scales of thousands of kilometers in the oceans. Provided that high-quality metagenomic assemblies and genomic bins or appropriate reference genomes are available, proteomics data can be resolved at fine taxonomic scales – at the level of strains – and can even be used to infer evolutionary processes such as patterns of genomic recombination (Denef et al. 2009, 2010b; Lo et al. 2007). In addition to the shotgun proteomics approaches that we focus on here, functional proteomics, including the in-gel identification of stained or active proteins, offers an effective way to link proteins with geomicrobiological processes (Dick et al. 2008; Yun et al. 2016).

10.1 Methodologies for Basic Proteomics

Proteomics includes several distinct approaches with varying levels of throughput. At the lower end of the throughput spectrum, proteins are first separated by one- or two-dimensional gel electrophoresis (Shevchenko et al. 1996; Wilmes & Bond 2004). Proteins are then extracted from the gel, digested with proteases (typically trypsin) to generate peptides, then analyzed by tandem mass spectrometry (MS/MS) to identify the peptides. More recently, advances in mass spectrometry have enabled high-throughput methods that can be applied directly to complex mixtures of peptides generated through direct proteolytic digestion of samples (Verberkmoes et al. 2009). The development of this "shotgun sequencing" approach parallels the evolution of metagenomic methods, which started with targeted approaches but shifted towards shotgun methods as throughput increased and costs decreased.

An overview of a shotgun proteomic workflow is shown in Figure 10.1. Following collection and preservation (Saito et al. 2011b), cells are lysed, typically by french press or sonication. To obtain information about the

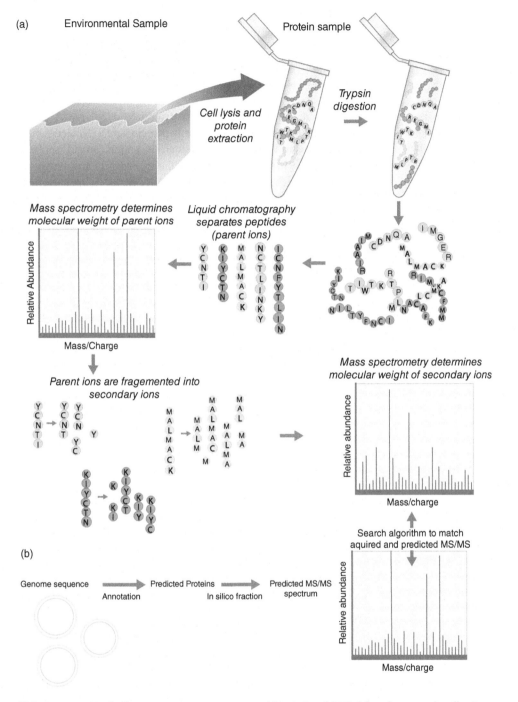

Figure 10.1 An overview of a shotgun proteomic workflow. Proteins are symbolized by chains of circles, with each circle representing an amino acid with its one-letter abbreviation. (a) Workflow from sample collection to mass spectrometry. (b) Bioinformatic generation of databases from genomic data. See text for details.

cellular location of each protein and maximize protein detection, the crude protein extract may be physically separated into membrane, cytoplasmic, and extracellular fractions (Ram et al. 2005). Proteins are extracted from the community and cleaved at sequence-specific sites with proteases such as trypsin, which cleaves at arginine and lysine. A detailed protocol for cell lysis, protein extraction, and trypsin digestion from seawater is provided in both written and video forms (Colatriano & Walsh 2015). This method has the advantage of providing both protein and DNA, enabling parallel metagenomic and metaproteomic analysis, which can aid in identification of proteins (see below). In this particular case, gel electrophoresis is used to eliminate SDS, which is used in the protein extraction but can affect trypsin digestion and mass spectrometry, and to fractionate the sample to improve proteomic coverage. Protein extraction from soils can suffer from interferences from elements such as humic substances, so the method of extraction can be critical for optimizing protein recovery (Keiblinger et al. 2012).

The resulting short peptides are then separated by nano-liquid chromatography, followed by a first iteration of mass spectrometry. These "parent ions" are then physically fragmented into a series of daughter ions that represent portions of the parent ion structure. This mixture of daughter ions is then run through a second iteration of mass spectrometry, and the resulting MS/MS spectra are matched against theoretical masses of parent and secondary ions predicted from genomic databases. Several search algorithms are available to perform this computational step, with SEQUEST (Eng et al. 2008) and MASCOT (Koenig et al. 2008) being two of the most widely used.

10.2   The Importance of Genomic Databases for Interpreting Proteomics Data

A key limitation of shotgun metaproteomics is the reliance on genomic databases for protein identification. *De novo* interpretation of MS/MS spectra without an underlying genomic database remains elusive, though progress in this area continues (Muth et al. 2013). Because even a single amino acid difference between target and reference database typically confounds peptide identification, the use of reference genomes that are substantially different from the target organisms is challenging. However, as sequences diverge, some peptide sequences will be conserved; since as few as two peptides can be used for identification of a protein, some sequence divergence between reference and target can be accommodated. Computational and experimental evidence shows that ~10% amino acid sequence divergence causes loss of half of identifiable proteins (Verberkmoes et al. 2009). Furthermore, since label-free quantitative approaches typically rely on the number of spectral matches per protein, lack of detection of peptides affects quantification efforts. Hence, the use of divergent genomes as references to interpret proteomic data has even more limitations than for metagenomics, where the limitations are substantial.

Proteogenomic approaches are only as powerful as the underlying genomic database, especially with regard to quantification and resolution of different microbial strains. A recent study shows that database selection is critical, and offers best practices for metaproteomic data interpretation and annotation (Timmins-Schiffman et al. 2017). Ideally, genomics and proteomics data is obtained from the same environmental sample. Where good genomic databases are available, such as for abundant marine cyanobacteria, proteomics can resolve numerous microbial species and even ecotypes (Saito et al. 2014, 2015). In addition to the challenge of identification of proteins, proteomics also inherits the challenge of functional annotation from genomics. Analysis of proteomic data in the context of operons (Ram et al. 2005) and pathways (Mosier et al. 2015) can help to link proteins to functions and to more effectively assess differential expression, respectively.

10.3 Quantitative Proteomics

Quantification of shotgun proteomic data is challenging for a number of reasons. First, the detectability of peptides during mass spectrometry is highly nonuniform (Jarnuczak et al. 2016). Second, some peptides may be preferentially lost due to processes such as adherence of hydrophobic peptides to surfaces. Finally, as with shotgun metagenomics and metatranscriptomics, proteomic approaches typically produce data that is in terms of relative abundance rather than absolute abundance. Despite these challenges, three major strategies for quantitative proteomics have been developed and are discussed below: label free, isotopic labeling, and isobaric labeling.

Quantitative proteomics without labeling must rely on metrics intrinsic to MS measurements, such as the intensity or area of peptide peaks (Old et al. 2005), spectral counts (Liu et al. 2004), and normalized spectral abundance factors (Florens et al. 2006). These approaches have been effectively applied to various microbial communities (Denef et al. 2010a; Hawley et al. 2014; Justice et al. 2012; Lauro et al. 2011; Lo et al. 2007; Morris et al. 2010; Ram et al. 2005; Schneider et al. 2012; Sowell et al. 2011). However, label-free methods of proteomic quantification are in terms of relative abundance, and even then their accuracy and precision are limited by the factors described above as well as run-to-run variation of mass spectrometry.

Stable isotope labeling is an effective method for obtaining more accurate proteomic quantification, and potentially absolute abundance. This includes a variety of approaches, with either *in vitro* or *in vivo* labeling, some targeted and others high throughput (or "global"), and also combinations of these two. The *in vivo* high-throughput approach that has been successfully applied to acid mine drainage microbial communities requires that microbial communities are grown with a source of isotopically labeled nutrient, typically ^{15}N, so that it is incorporated into every N atom of every protein (Pan & Banfield 2014). This labeled "reference" sample is then mixed in equal proportions with an unlabeled, "unknown" field sample and the mixture is run on LC-MS/MS. The relative intensity of the mass spectra provides the relative abundance of known versus unknown proteins (Belnap et al. 2010, 2011). A major disadvantage of *in vivo* isotopic labeling for proteomics is that the cultures or communities need to be grown in the laboratory in order for the stable isotopes to be effectively incorporated.

Absolute proteomic quantification can be achieved by using synthetic isotopically labeled peptide standards (with deuterium, ^{13}C, or ^{15}N) for proteins of interest (Barnidge et al. 2003; Gerber et al. 2003). This targeted approach and the resulting absolute abundances can then be related to the relative abundance data provided by LC-MS parent ion spectra. If this is done with a sufficient number of isotopically labeled peptide standards for different proteins, a calibration for absolute quantification of unknown peptides (i.e., in terms of moles or grams of protein per liter) can be constructed (Malmstrom et al. 2009). The main disadvantage of this approach is in terms of throughput and cost. Nevertheless, it is effective for scientific questions in which certain proteins can be used as markers of biogeochemical processes (Saito et al. 2015). A combination of global and targeted quantitative proteomics has proven to be valuable in studying marine cyanobacterial communities and their response to nutrient availability (Saito et al. 2011a, 2014).

Most recently, proteomic quantification has been conducted by chemical isobaric labeling of peptides in a trypsin-digested mixture *in vitro*: Tandem Mass Tag (TMT) and Isobaric Tag for Relative and Absolute Quantification (iTRAQ) (Rauniyar & Yates 2014). "Isobaric" in this case means that the same overall mass is added to peptides from multiple samples, but those tags fragment during MS/MS analysis, yielding multiple "reporter ions" whose intensities are proportional to the abundance of a peptide in each sample. These methods offer improved accuracy over label-free methods of quantification (Li et al. 2012). Peptides from different samples are labeled differently and then pooled and run on the same LC-MS/MS run, thus avoiding concerns about run-to-run variation (though variation due to sample preparation still needs to be considered). An example application of the TMT approach to microbial communities was to study the impact of temperature on individual microbial groups within acid mine drainage (Mosier et al. 2015).

These isobaric labeling approaches can simultaneously quantify many hundreds or even thousands of proteins, and yield results in terms of relative abundance (i.e., what is the abundance of protein X in sample A versus sample B?). Where absolute quantification is required, for example comparing the abundance of two different proteins in a single sample or placing protein quantities in terms of moles or grams per volume of sample, targeted proteomics with individual synthetic standards is required.

10.4 Combining Stable Isotope Probing with Proteomics to Track Microbial Metabolism

Stable isotope probing can also be used to identify metabolically active microbes, to determine which organisms are incorporating isotopically labeled (^{13}C, ^{15}N, and ^{36}S) substrates into protein, and to track metabolic interactions between community members via cross-feeding (von Bergen et al. 2013). "Metabolic labeling" of proteins can be done with isotopically labeled amino acids that are then directly incorporated into the protein, or by isotopically labeled nutrients or substrates that are of interest, such as inorganic ($H^{13}CO_3^-$, $^{15}NO_3^-$) or organic (^{13}C-acetate, ^{15}N-urea) carbon and nitrogen sources. Direct incorporation yields signature spectral distributions and can be distinguished from indirect incorporation via subsequent release and uptake of metabolites or degradation products (von Bergen et al. 2013). Hence, SIP-proteomics can be used to trace the flux of elements through microbial communities via metabolic interactions, such as between autotrophs and heterotrophs (Justice et al. 2014). This method has also been used to determine which community members are responsible for the degradation of various organic contaminants and hydrocarbons (Bastida et al. 2009; Bozinovski et al. 2012; Jechalke et al. 2013; Morris et al. 2012; Taubert et al. 2012).

Finally, proteomics can also be used to explore natural isotopic "fingerprints" of microbial communities as they exist in the environment. Mohr et al. (2014) have developed such a protein stable isotope fingerprinting (P-SIF) method, which measures the carbon isotope values (δ^{13}C) of proteins in order to link taxonomic identity of microbes to metabolic function. While to date, this method has only been applied in a proof-of-concept study involving two species, it offers great promise in linking particular species to major metabolisms that impart isotopic fractionation (methanogenesis, methanotrophy, different pathways of autotrophy), and possibly for resolving trophic interactions.

References

Abraham, P. E., Giannone, R. J., Xiong, W. & Hettich, R. L. (2014) Metaproteomics: extracting and mining proteome information to characterize metabolic activities in microbial communities. *Current Protocols in Bioinformatics*, **46**, 13.26.1–14.

Barnidge, D. R., Dratz, E. A., Martin, T., Bonilla, L. E., Moran, L. B. & Lindall, A. (2003. Absolute quantification of the G protein-coupled receptor rhodopsin by LC/MS/MS using proteolysis product peptides and synthetic peptide standards. *Analytical Chemistry*, **75**, 445–451.

Bastida, F., Moreno, J. L., Nicolas, C., Hernandez, T. & Garcia, C. (2009) Soil metaproteomics: a review of an emerging environmental science. Significance, methodology and perspectives. *European Journal of Soil Science*, **60**, 845–859.

Belnap, C. P., Pan, C., Verberkmoes, N. C., et al. (2010) Cultivation and quantitative proteomic analyses of acidophilic microbial communities. *ISME Journal*, **4**, 520–530.

Belnap, C. P., Pan, C., Denef, V. J., Samatova, N. F., Hettich, R. L. & Banfield, J. F. (2011) Quantitative proteomic analyses of the response of acidophilic microbial communities to different pH conditions. *ISME Journal*, **5**, 1152–1161.

Bozinovski, D., Herrmann, S., Richnow, H. H., von Bergen, M., Seifert, J. & Vogt, C. (2012) Functional analysis of an anaerobic m-xylene-degrading enrichment culture using protein-based stable isotope probing. *FEMS Microbiology Ecology*, **81**, 134–144.

Colatriano, D. & Walsh, D. A. (2015) An aquatic microbial metaproteomics workflow: from cells to tryptic peptides suitable for tandem mass spectrometry-based analysis. *Journal of Visualized Experiments*, 103, 10.3791/52827.

Denef, V. J., Verberkmoes, N. C., Shah, M. B., Abraham, P., Lefsrud, M., Hettich, R. L. & Banfield, J. F. (2009) Proteomics-inferred genome typing (PIGT) demonstrates inter-population recombination as a strategy for environmental adaptation. *Environmental Microbiology*, **11**, 313–325.

Denef, V. J., Kalnejais, L. H., Mueller, R. S., et al. (2010a) Proteogenomic basis for ecological divergence of closely related bacteria in natural acidophilic microbial communities. *Proceedings of the National Academy of Sciences of the United States of America*, **107**, 2383–2390.

Denef, V. J., Mueller, R. S. & Banfield, J. F. (2010b. AMD biofilms: using model communities to study microbial evolution and ecological complexity in nature. *Isme Journal*, **4**, 599–610.

Dick, G. J., Torpey, J. W., Beveridge, T. J. & Tebo, B. M. (2008) Direct identification of a bacterial manganese(II) oxidase, the multicopper oxidase MnxG, from spores of several different marine *Bacillus* species. *Applied and Environmental Microbiology*, **74**, 1527–1534.

Eng, J. K., Fischer, B., Grossmann, J. & Maccoss, M. J. (2008) A fast SEQUEST cross correlation algorithm. *Journal of Proteome Research*, **7**, 4598–4602.

Florens, L., Carozza, M. J., Swanson, S. K., et al. (2006) Analyzing chromatin remodeling complexes using shotgun proteomics and normalized spectral abundance factors. *Methods*, **40**, 303–311.

Gerber, S. A., Rush, J., Stemman, O., Kirschner, M. W. & Gygi, S. P. (2003) Absolute quantification of proteins and phosphoproteins from cell lysates by tandem MS. *Proceedings of the National Academy of Sciences of the United States of America*, **100**, 6940–6945.

Hawley, A. K., Brewer, H. M., Norbeck, A. D., Pasa-Tolic, L. & Hallam, S. J. (2014) Metaproteomics reveals differential modes of metabolic coupling among ubiquitous oxygen minimum zone microbes. *Proceedings of the National Academy of Sciences of the United States of America*, **111**, 11395–11400.

Jaffe, J. D., Berg, H. C. & Church, G. M. (2004) Proteogenomic mapping as a complementary method to perform genome annotation. *Proteomics*, **4**, 59–77.

Jarnuczak, A. F., Lee, D. C., Lawless, C., Holman, S. W., Eyers, C. E. & Hubbard, S. J. (2016) Analysis of intrinsic peptide detectability via integrated label-free and SRM-based absolute quantitative proteomics. *Journal of Proteome Research*, **15**, 2945–2959.

Jechalke, S., Franchini, A. G., Bastida, F., et al. (2013) Analysis of structure, function, and activity of a benzene-degrading microbial community. *FEMS Microbiology Ecology*, **85**, 14–26.

Justice, N. B., Pan, C., Mueller, R., et al. (2012) Heterotrophic archaea contribute to carbon cycling in low-pH, suboxic biofilm communities. *Applied and Environmental Microbiology*, **78**, 8321–8330.

Justice, N. B., Li, Z., Wang, Y., Spaudling, S. E., et al. (2014) (15)N- and (2)H proteomic stable isotope probing links nitrogen flow to archaeal heterotrophic activity. *Environmental Microbiology*, **16**, 3224–3237.

Keiblinger, K. M., Wilhartitz, I. C., Schneider, T., et al. (2012) Soil metaproteomics – comparative evaluation of protein extraction protocols. *Soil Biology and Biochemistry*, **54**, 14–24.

Koenig, T., Menze, B. H., Kirchner, M., et al. (2008) Robust prediction of the MASCOT score for an improved quality assessment in mass spectrometric proteomics. *Journal of Proteome Research*, **7**, 3708–3717.

Lauro, F. M., Demaere, M. Z., Yau, S., et al. (2011) An integrative study of a meromictic lake ecosystem in Antarctica. *ISME Journal*, **5**, 879–895.

Li, Z., Adams, R. M., Chourey, K., Hurst, G. B., Hettich, R. L. & Pan, C. (2012) Systematic comparison of label-free, metabolic labeling, and isobaric chemical labeling for quantitative proteomics on LTQ Orbitrap Velos. *Journal of Proteome Research*, **11**, 1582–1590.

Liu, H., Sadygov, R. G. & Yates, J. R. 3rd (2004) A model for random sampling and estimation of relative protein abundance in shotgun proteomics. *Analytical Chemistry*, **76**, 4193–4201.

Lo, I., Denef, V. J., Verberkmoes, N. C., et al. (2007) Strain-resolved community proteomics reveals recombining genomes of acidophilic bacteria. *Nature*, **446**, 537–541.

Malmstrom, J., Beck, M., Schmidt, A., Lange, V., Deutsch, E. W. & Aebersold, R. (2009) Proteome-wide cellular protein concentrations of the human pathogen Leptospira interrogans. *Nature*, **460**, 762–765.

Mohr, W., Tang, T., Sattin, S. R., Bovee, R. J. & Pearson, A. (2014) Protein stable isotope fingerprinting: multidimensional protein chromatography coupled to stable isotope–ratio mass spectrometry. *Analytical Chemistry*, **86**, 8514–8520.

Morris, B. E., Herbst, F. A., Bastida, F., et al. (2012) Microbial interactions during residual oil and n-fatty acid metabolism by a methanogenic consortium. *Environmental Microbiology Reports*, **4**, 297–306.

Morris, R. M., Nunn, B. L., Frazar, C., Goodlett, D. R., Ting, Y. S. & Rocap, G. (2010) Comparative metaproteomics reveals ocean-scale shifts in microbial nutrient utilization and energy transduction. *ISME Journal*, **4**, 673–685.

Mosier, A. C., Li, Z., Thomas, B. C., Hettich, R. L., Pan, C. & Banfield, J. F. (2015) Elevated temperature alters proteomic responses of individual organisms within a biofilm community. *ISME Journal*, **9**, 180–194.

Muth, T., Benndorf, D., Reichl, U., Rapp, E. & Martens, L. (2013) Searching for a needle in a stack of needles: challenges in metaproteomics data analysis. *Molecular Biosystems*, **9**, 578–585.

Old, W. M., Meyer-Arendt, K., Aveline-Wolf, L., et al. (2005) Comparison of label-free methods for quantifying human proteins by shotgun proteomics. *Molecular and Cellular Proteomics*, **4**, 1487–1502.

Pan, C. & Banfield, J. F. (2014) Quantitative metaproteomics: functional insights into microbial communities. In: I. T. Paulsen & A. J. Holmes (eds), *Environmental Microbiology Methods and Protocols*, 2nd edn. Humana Press, New York.

Ram, R. J., Verberkmoes, C., Thelen, M. P., et al. (2005) Community Proteomics of a natural microbial biofilm. *Science*, **308**, 1915–1920.

Rauniyar, N. & Yates, J. R. 3rd (2014) Isobaric labeling-based relative quantification in shotgun proteomics. *Journal of Proteome Research*, **13**, 5293–5309.

Saito, M. A., Bertrand, E. M., Dutkiewicz, S., et al. (2011a) Iron conservation by reduction of metalloenzyme inventories in the marine diazotroph Crocosphaera watsonii. *Proceedings of the National Academy of Sciences of the United States of America*, **108**, 2184–2189.

Saito, M. A., Bulygin, V. V., Moran, D. M., Taylor, C. & Scholin, C. (2011b) Examination of microbial proteome preservation techniques applicable to autonomous environmental sample collection. *Frontiers in Microbiology*, **2**, 215.

Saito, M. A., McIlvin, M. R., Moran, D. M., et al. (2014) Multiple nutrient stresses at intersecting Pacific Ocean biomes detected by protein biomarkers. *Science*, **345**, 1173–1177.

Saito, M. A., Dorsk, A., Post, A. F., et al. (2015) Needles in the blue sea: sub-species specificity in targeted protein biomarker analyses within the vast oceanic microbial metaproteome. *Proteomics*, **15**, 3521–3531.

Schneider, T., Keiblinger, K. M., Schmid, E., et al. (2012) Who is who in litter decomposition? Metaproteomics reveals major microbial players and their biogeochemical functions. *ISME Journal*, **6**, 1749–1762.

Shevchenko, A., Jensen, O. N., Podtelejnikov, A. V., et al. (1996) Linking genome and proteome by mass spectrometry: large-scale identification of yeast proteins from two dimensional gels. *Proceedings of the National Academy of Sciences of the United States of America*, **93**, 14440–14445.

Singer., S. W., Chan, C. S., Zemla, A., et al. (2008) Characterization of cytochrome 579, an unusual cytochrome isolated from an iron-oxidizing microbial community. *Applied and Environmental Microbiology*, **74**, 4454–4462.

Sowell, S. M., Abraham, P. E., Shah, M., et al. (2011) Environmental proteomics of microbial plankton in a highly productive coastal upwelling system. *ISME Journal*, **5**, 856–865.

Taubert, M., Vogt, C., Wubet, T., et al. (2012) Protein-SIP enables time-resolved analysis of the carbon flux in a sulfate-reducing, benzene-degrading microbial consortium. *ISME Journal*, **6**, 2291–2301.

Timmins-Schiffman, E., May, D. H., Mikan, M., et al. (2017) Critical decisions in metaproteomics: achieving high confidence protein annotations in a sea of unknowns. *ISME Journal*, **11**, 309–314.

Tyson, G. W., Chapman, J., Hugenholtz, P., et al. (2004) Community structure and metabolism through reconstruction of microbial genomes from the environment. *Nature*, **428**, 37–43.

Verberkmoes, N. C., Denef, V. J., Hettich, R. L. & Banfield, J. F. (2009) Systems biology: functional analysis of natural microbial consortia using community proteomics. *Nature Reviews Microbiology*, **7**, 196–205.

Von Bergen, M., Jehmlich, N., Taubert, M., et al. (2013) Insights from quantitative metaproteomics and protein-stable isotope probing into microbial ecology. *ISME Journal*, **7**, 1877–1885.

Wilmes, P. & Bond, P. L. (2004) The application of two-dimensional polyacrylamide gel electrophoresis and downstream analyses to a mixed community of prokaryotic microorganisms. *Environmental Microbiology*, **6**, 911–920.

Yun, J., Malvankar, N. S., Ueki, T. & Lovley, D. R. (2016) Functional environmental proteomics: elucidating the role of a c-type cytochrome abundant during uranium bioremediation. *ISME Journal*, **10**, 310–320.

11 Lipidomics and Metabolomics

Introduction

While this book focuses on methods for extracting information from the DNA, RNA, and proteins of microbial cells and communities, lipids and metabolites also provide unique and valuable molecular information about microbial geochemistry. Because of their exceptional potential for preservation in sediments and sedimentary rocks, lipids are especially valuable in the context of the geological record, where they can provide insights into the ecosystems and environments that characterized Earth millions and even billions of years ago (Briggs & Summons 2014; Brocks & Banfield 2009; Brocks & Pearson 2005; Newman et al. 2016; Summons & Lincoln 2012). They also may carry stable isotopic signatures that are indicative of certain metabolisms or utilization of specific substrates, such as methane. In the context of microbial geochemistry, lipidomics refers to the study of lipids and the genes and enzymes involved in their biosynthesis as they relate to ecosystem function and Earth history (Pearson 2014). Likewise, metabolomics refers to the study of the pool of small-molecule metabolites associated with microbial cells or communities, which reflects the integrated activities of metabolic pathways and processes and the expression of their underlying genes and proteins (Moran et al. 2016).

11.1 Lipidomics

Unfortunately, the functional and taxonomic information preserved within lipids usually pales in comparison to that held within DNA, RNA, or protein sequences. Some lipid biomarkers are diagnostic of archaea, eukarya, and

Genomic Approaches in Earth and Environmental Sciences, First Edition. Gregory Dick.
© 2019 John Wiley & Sons Ltd. Published 2019 by John Wiley & Sons Ltd.

bacteria (Briggs & Summons 2014; Summons & Lincoln 2012), but there are relatively few cases where specific lipids have been conclusively linked to specific taxonomic groups, such as the association of ladderanes with anammox Planctomycetes (Pearson 2014; Sinninghe Damste et al. 2002). However, these organic biomarkers are often faithful indicators of specific metabolic or biosynthetic pathways, and the resolution and confidence of the information can be enhanced with isotopic data from the organic compounds (Brocks & Pearson 2005; Hayes 2001; Hinrichs et al. 2001; Zhang et al. 2009). Some lipid biomarkers may also reflect physiological state (Welander & Summons 2012). Further, the structure of some lipids correlates systematically with environmental temperature and thus can provide insights into paleoclimate (Eglinton & Eglinton 2008; Kim et al. 2010). Hence, despite their limitations, lipids are an indispensable component of the geobiological toolbox due to their high preservation potential, which provides opportunities to study organisms and ecosystems from millions and even billions of years ago.

Steranes such as cholestane have been taken as evidence for early eukaryotes (Brocks et al. 1999), although recent advances have shown that reports of Archean steranes are artifacts of contamination (French et al. 2015). More rigorous contamination controls imposed reveal a revised biomarker record and suggest that eukaryotic algae did not rise to ecological prominence until the Neoproterozoic (Brocks et al. 2017). Hence, even where phylogenetic resolution is low, lipid biomarkers can provide incredibly valuable insights. Steranes can also yield information about ancient metabolisms and environments. Because steroid biosynthesis requires molecular O_2, steranes in the geological record provide a unique marker of O_2, and knowledge of just how much O_2 is required by biosynthetic pathways can be used to quantitatively constrain ancient O_2 concentrations (Waldbauer et al. 2011). The molecular biomarker record of plant and animal life has also proven to be invaluable, providing insights into the rise of major groups such as diatoms (Sinninghe Damste et al. 2004), as well as characteristics of particular fossils such as the original color of various tissues (Briggs & Summons 2014).

Efforts to map the distribution of biomarkers across the tree of life now find new opportunities in the midst of the genome sequencing revolution (Brocks & Pearson 2005; Brocks & Banfield 2009; Newman et al. 2016; Summons et al. 2006). Biosynthetic pathways for biomarkers of interest can be quickly found with computer searches of genome sequences (Pearson et al. 2003). However, this requires an understanding of the links between biosynthetic genes and biomarkers, which is a formidable challenge in itself. Advances in high-throughput functional approaches can uncover such linkages, and are discussed in depth with examples in Pearson (2014). Recent studies illustrate the crucial need to understand how specific compounds are distributed across the tree of life in order to successfully interpret the biomarker record. Whereas 2-methylhopanes were originally considered as a marker of oxygenic photosynthesis (Brocks et al. 1999), subsequent studies showed that they can also be produced in abundance by other organisms that do not conduct oxygenic photosynthesis (Rashby et al. 2007; Welander et al. 2010). Newman et al. (2016) synthesize this story to illustrate how cellular and molecular biological approaches in modern systems are central to the interpretation of biomarkers in the rock record. In a more recent example, the wider phylogenetic distribution of biosynthetic pathways for tetrahymanol, a precursor of the diagenetic product gammacerane, raised questions about its common use as a biomarker of water column stratification (Banta et al. 2015, 2017).

Advances in high-throughput production of lipid data open exciting new opportunities for investigating lipid biomarkers in the sedimentary and geological record. Direct detection of target lipids in sediment sections at fine spatial scales enables the use of lipid biomarkers to track records of climate change on subannual to decadal time scales (Wörmer et al. 2014), and advanced statistical techniques are increasingly being employed to help understand the biomarker data (Tierney & Tingley 2015). When combined with a better understanding of the phylogenetic and physiological meaning of specific lipids derived from the current revolution in DNA sequencing, proteomics, and other high-throughput functional approaches (Brocks & Banfield 2009; Pearson 2014), the future of lipidomics in geobiology looks particularly promising.

11.2 Metabolomics

Metabolomics is the study of the complete set of small molecules (metabolites) within a microbial culture or community. It provides a valuable snapshot of the physiological status of microorganisms (Kido Soule et al. 2015; Tang 2011). Like lipidomics, metabolomics lags behind genomics, transcriptomics, and proteomics, particularly in the context of whole microbial communities, where there have been comparatively few studies. In part, this is due to experimental challenges associated with identifying metabolites in complex environmental matrices. However, these challenges can be overcome to some extent by advances in analytical (Johnson et al. 2017; Kido Soule et al. 2015), computational (Longnecker & Kujawinski 2017; Longnecker et al. 2015), and stable isotope labeling approaches (Mosier et al. 2013). The upside is huge because metabolomics can provide important insights into the metabolic state of microorganisms, and how different microbes interact with each other and with their surroundings.

Metabolomics is particularly powerful when done in parallel with proteomics because together they can produce a picture of what metabolites are present and what pathways are actively involved in their production, cellular uptake, and utilization. For example, Halter et al. (2012) characterized metabolites secreted by a photosynthetic protist in acid mine drainage, identifying specific compounds, and their corresponding proteins of biosynthetic pathways, that potentially impact the associated microbial community. Also in acid mine drainage, Mosier et al. (2013) identified key metabolites for adaptation to acidic, metal-rich conditions. Metagenomic data can also be used to make predictions about which taxa are potentially responsible for production and consumption of metabolites (Larsen et al. 2011).

Results of metabolomics studies conducted in pure cultures lay a foundation for interpreting metabolomic studies of microbial communities. Steffen and colleagues found that the metabolome of the cyanobacterium *Microcystis aeruginosa* was strikingly static compared to the transcriptome, and inferred that these two omics approaches are sensitive to different levels of environmental change (Steffen et al. 2014). These results emphasize that the mRNA pool turns over more quickly than the metabolite pool, and transcriptomic and proteomic responses may be mounted to stabilize the metabolome. However, in other cases metabolomics has been used to identify small molecules potentially involved in adaptation to environmental changes in conditions such as temperature (Ghobakhlou et al. 2013; Trauger et al. 2008).

Metabolomics is also valuable in studying the interactions between microorganisms. It has been used to identify cell-to-cell signaling pathways between cyanobacteria and associated heterotrophs that are involved in regulating nutrient uptake (van Mooy et al. 2012). Viral infections promote dramatic changes in the metabolomes of infected populations, with implications for biogeochemical effects of virally mediated lysis (Ankrah et al. 2014). In some or perhaps even most cases, the exchange of compounds between organisms can be cryptic, due to low concentrations, rapid turnover, unknown compounds, or other analytical challenges.

New computational approaches show great promise in identification of environmentally relevant metabolites (Longnecker & Kujawinski 2017). Here, the utilization of transcriptomic approaches can be valuable, as was demonstrated in a co-culture of a marine autotroph and heterotroph (Durham et al. 2015).

Overall, these studies show how multiple omics approaches can be employed in parallel to draw on the specific strengths and minimize the weaknesses of each.

References

Ankrah, N. Y., May, A. L., Middleton, J. L., et al. (2014) Phage infection of an environmentally relevant marine bacterium alters host metabolism and lysate composition. *ISME Journal*, **8**, 1089–1100.

Banta, A. B., Wei, J. H. & Welander, P. V. (2015) A distinct pathway for tetrahymanol synthesis in bacteria. *Proceedings of the National Academy of Sciences of the United States of America*, **112**, 13478–13483.

Banta, A. B., Wei, J. H., Gill, C. C., Giner, J. L. & Welander, P. V. (2017) Synthesis of arborane triterpenols by a bacterial oxidosqualene cyclase. *Proceedings of the National Academy of Sciences of the United States of America*, **114**, 245–250.

Briggs, D. E. & Summons, R. E. (2014) Ancient biomolecules: their origins, fossilization, and role in revealing the history of life. *Bioessays*, **36**, 482–490.

Brocks, J. J. & Banfield, J. (2009) Unravelling ancient microbial history with community proteogenomics and lipid geochemistry. *Nature Reviews Microbiology*, **7**, 601–609.

Brocks, J. J. & Pearson, A. (2005) Building the biomarker tree of life. *Molecular Geomicrobiology*, **59**, 233–258.

Brocks, J. J., Logan, G. A., Buick, R. & Summons, R. E. (1999) Archean molecular fossils and the early rise of eukaryotes. *Science*, **285**, 1033–1036.

Brocks, J. J., Jarrett, A. J. M., Sirantoine, E., Hallmann, C., Hoshino, Y. & Liyanage, T. (2017) The rise of algae in Cryogenian oceans and the emergence of animals. *Nature*, **548**, 578–581.

Durham, B. P., Sharma, S., Luo, H., et al. (2015) Cryptic carbon and sulfur cycling between surface ocean plankton. *Proceedings of the National Academy of Sciences of the United States of America*, **112**, 453–457.

Eglinton, T. I. & Eglinton, G. (2008) Molecular proxies for paleoclimatology. *Earth and Planetary Science Letters*, **275**, 1–16.

French, K. L., Hallmann, C., Hope, J. M., et al. (2015) Reappraisal of hydrocarbon biomarkers in Archean rocks. *Proceedings of the National Academy of Sciences of the United States of America*, **112**, 5915–5920.

Ghobakhlou, A., Laberge, S., Antoun, H., et al. (2013) Metabolomic analysis of cold acclimation of Arctic Mesorhizobium sp. strain N33. *PLoS One*, **8**, e84801.

Halter, D., Goulhen-Chollet, F., Gallien, S., et al. (2012) In situ proteo-metabolomics reveals metabolite secretion by the acid mine drainage bio-indicator, Euglena mutabilis. *ISME Journal*, **6**, 1391–1402.

Hayes, J. M. (2001) Fractionation of carbon and hydrogen isotopes in biosynthetic processes. *Reviews in Mineralogy and Geochemistry*, **43**, 225–277.

Hinrichs, K. U., Eglinton, G., Engel, M. H. & Summons, R. E. (2001) Exploiting the multivariate isotopic nature of organic compounds. *Geochemistry Geophysics Geosystems*, **2**, art. no. 2000GC000142.

Johnson, W. M., Kido Soule, M. C. & Kujawinski, E. B. (2017) Extraction efficiency and quantification of dissolved metabolites in targeted marine metabolomics. *Limnology and Oceanography: Methods*, **15**, 417–428.

Kido Soule, M. C., Longnecker, K., Johnson, W. M. & Kujawinski, E. B. (2015) Environmental metabolomics: analytical strategies. *Marine Chemistry*, **177**, 374–387.

Kim, J. H., van der Meer, J., Schouten, S., et al. (2010) New indices and calibrations derived from the distribution of crenarchaeal isoprenoid tetraether lipids: Implications for past sea surface temperature reconstructions. *Geochimica et Cosmochimica Acta*, **74**, 4639–4654.

Larsen, P. E., Collart, F. R., Field, D., et al. (2011) Predicted Relative Metabolomic Turnover (PRMT): determining metabolic turnover from a coastal marine metagenomic dataset. *Microbial Informatics and Experimentation*, **1**, 4.

Longnecker, K. & Kujawinski, E. B. (2017) Mining mass spectrometry data: using new computational tools to find novel organic compounds in complex environmental mixtures. *Organic Geochemistry*, **110**, 92–99.

Longnecker, K., Futrelle, J., Coburn, E., Kido Soule, M. C. & Kujawinski, E. B. (2015) Environmental metabolomics: databases and tools for data analysis. *Marine Chemistry*, **177**, 366–373.

Moran, M. A., Kujawinski, E. B., Stubbins, A., et al. (2016) Deciphering ocean carbon in a changing world. *Proceedings of the National Academy of Sciences of the United States of America*, **113**, 3143–3151.

Mosier, A. C., Justice, N. B., Bowen, B. P., et al. (2013) Metabolites associated with adaptation of microorganisms to an acidophilic, metal-rich environment identified by stable-isotope-enabled metabolomics. *MBio*, **4**, e00484-12.

Newman, D. K., Neubauer, C., Ricci, J. N., Wu, C. H. & Pearson, A. (2016) Cellular and molecular biological approaches to interpreting ancient biomarkers. *Annual Review of Earth and Planetary Sciences*, **44**, 493–522.

Pearson, A. (2014) Lipidomics for geochemistry. In: *Treatise on Geochemistry, vol. 12, Organic Geochemistry*, 2nd edn, pp. 291–336.

Pearson, A., Budin, M. & Brocks, J. J. (2003) Phylogenetic and biochemical evidence for sterol synthesis in the bacterium Gemmata obscuriglobus. *Proceedings of the National Academy of Sciences of the United States of America*, **100**, 15352–15357.

Rashby, S. E., Sessions, A. L., Summons, R. E. & Newman, D. K. (2007) Biosynthesis of 2-methylbacteriohopanepolyols by an anoxygenic phototroph. *Proceedings of the National Academy of Sciences of the United States of America*, **104**, 15099–15104.

Sinninghe Damste, J. S., Strous, M., Rijpstra, W. I. C., et al. (2002) Linearly concatenated cyclobutane lipids from a dense bacterial membrane. *Nature*, **419**, 708–712.

Sinninghe Damste, J. S., Muyzer, G., Abbas, B., et al. (2004) The rise of the rhizosolenoid diatoms. *Science*, **304**, 584–587.

Steffen, M. M., Dearth, S. P., Dill, B. D., et al. (2014) Nutrients drive transcriptional changes that maintain metabolic homeostasis but alter genome architecture in Microcystis. *ISME Journal*, **8**, 2080–2092.

Summons, R. E. & Lincoln, S. A. (2012) Biomarkers: informative molecules for studies in geobiology. In: A. H. Knoll, D. E. Canfield & K. O. Konhauser (eds), *Fundamentals of Geobiology*. Wiley-Blackwell, Chichester.

Summons, R. E., Bradley, A. S., Jahnke, L. L. & Waldbauer, J. R. (2006) Steroids, triterpenoids and molecular oxygen. *Philosophical Transactions of the Royal Society of London B Biological Science*, **361**, 951–968.

Tang, J. (2011) Microbial metabolomics. *Current Genomics*, **12**, 391–403.

Tierney, J. E. & Tingley, M. P. (2015) A TEX$_{86}$ surface sediment database and extended Bayesian calibration. *Scientific Data*, **2**, 150029.

Trauger, S. A., Kalisak, E., Kalisiak, J., et al. (2008) Correlating the transcriptome, proteome, and metabolome in the environmental adaptation of a hyperthermophile. *Journal of Proteome Research*, **7**, 1027–1035.

Van Mooy, B. A., Hmelo, L. R., Sofen, L. E., et al. (2012) Quorum sensing control of phosphorus acquisition in Trichodesmium consortia. *ISME Journal*, **6**, 422–429.

Waldbauer, J. R., Newman, D. K. & Summons, R. E. (2011) Microaerobic steroid biosynthesis and the molecular fossil record of Archean life. *Proceedings of the National Academy of Sciences of the United States of America*, **108**, 13409–13414.

Welander, P. V. & Summons, R. E. (2012) Discovery, taxonomic distribution, and phenotypic characterization of a gene required for 3-methylhopanoid production. *Proceedings of the National Academy of Sciences of the United States of America*, **109**, 12905–12910.

Welander, P. V., Coleman, M. L., Sessions, A. L., Summons, R. E. & Newman, D. K. (2010) Identification of a methylase required for 2-methylhopanoid production and implications for the interpretation of sedimentary hopanes. *Proceedings of the National Academy of Sciences of the United States of America*, **107**, 8537–8542.

Wörmer, L., Elvert, M., Fuchser, J., et al. (2014) Ultra-high-resolution paleoenvironmental records via direct laser-based analysis of lipid biomarkers in sediment core samples. *Proceedings of the National Academy of Sciences of the United States of America*, **111**, 15669–15674.

Zhang, X., Gillespie, A. L. & Sessions, A. L. (2009) Large D/H variations in bacterial lipids reflect central metabolic pathways. *Proceedings of the National Academy of Sciences of the United States of America*, **106**, 12580–12586.

12

Downstream and Integrative Approaches and Future Outlook

Introduction

Whereas previous chapters focused on the "nuts and bolts" of specific omics approaches, this chapter covers downstream tasks and issues surrounding data products generated by earlier stages of analysis. Although ultimately critical for realizing the full potential of omics approaches for the Earth and environmental sciences, associated resources and infrastructure remain underdeveloped and insufficient in many cases. Nevertheless, progress is being made as more and more publicly available tools make analysis of omics datasets more accessible to the nonexpert. Efforts led by both the community and funding agencies are making progress towards establishment of infrastructure and standard practices for analyzing, storing, and sharing data. Recent advances in the integration of omics data into numerical models are particularly exciting because modeling offers vast potential for synthesizing large datasets and directly testing our level of understanding of microbial systems, and eventually for moving the field toward the ability to make predictions about outcomes of microbially mediated processes.

12.1 Comparative Omics

Whether derived from a culture, single cell, or a genomic bin of a community, comparing the gene content of different genomes is a central task in genomics. For example, defining which genes are unique (or common) to a certain phylogenetic or functional group of organisms can reveal genetic and functional themes and differences. The core challenge is usually to define

Genomic Approaches in Earth and Environmental Sciences, First Edition. Gregory Dick.
© 2019 John Wiley & Sons Ltd. Published 2019 by John Wiley & Sons Ltd.

orthologs, homologous genes that serve a conserved function (as opposed to paralogs, which are homologous genes derived from gene duplication that often have different functions). One method to do this is a pair-wise all versus all BLAST, where the reciprocal best hits above a certain threshold are defined as orthologs (Kristensen et al. 2011). Markov chain clustering (MCL) is an effective method of comparing the gene content of multiple genomes (Li et al. 2003). Several software packages are available to accomplish this task, including publicly available programs IMG (Markowitz et al. 2009) and Pogo-DB (Lan et al. 2014), which are useful where the genomes are publicly available and comparisons have been precomputed. GET_HOMOLOGUES can define homologs using several different methods and produce graphical output on components of the pangenome (Contreras-Moreira & Vinuesa 2013). Anvi'o also has a workflow for microbial pangenomics which can identify protein clusters and compare them across genomes and metagenome-assembled genomes (Eren 2017b).

Data derived from whole communities or from single cells presents several challenges for comparative analyses. First, for genome-centric comparisons, data derived from metagenomic data or single cells is often incomplete. Methods to compare partial genomes, such as the modified mediator genome reference assembly approach, have been developed (Kashtan et al. 2014; Wurtzel et al. 2010). Second, omics data from communities often comes in the form of relative abundance, not absolute abundance. Thus, as discussed in the section on metatranscriptomics, careful consideration must be given to strategies for normalization and statistical inference of differential abundance.

12.2   Statistical Approaches

As DNA sequencing costs have declined, it has become possible to conduct community omics studies that include replicate sequencing of conditions or samples. This enables the application of rigorous statistical approaches to microbial community omics data, which are required to identify biologically meaningful differences between microbial communities. Many statistical approaches and pipelines that are applicable to microbial ecology data, especially 16S rRNA gene data (McMurdie & Holmes 2013), are well established. However, fewer tools have been developed for shotgun omics data, and like many aspects of research related to next-generation sequencing, keeping up with the best practices and approaches can be daunting (Buttigieg & Ramette 2014a). Hence, software packages and dynamic online resources such as blogs, forums, and "how-to" guides are invaluable. The GUSTA ME (GUide to STatistical Analysis in Microbial Ecology) (Buttigieg & Ramette 2014b) and STAMPS (Strategies and Techniques for Analyzing Microbial Population Structure) (STAMPS 2017) websites (and course) provide valuable guides in this regard and contain some of the most common and useful tools for statistical analysis of community omics (Buttigieg & Ramette 2014a; Parks et al. 2014).

Also widely used are statistical packages within R (Ismay & Kim 2015), those for general ecological analyses such as VEGAN (Simpson et al. 2017) and PRIMER-E (PRIMER-E 2017), and those borrowed from analysis of differential expression RNAseq data on single organisms such as DESeq (Love et al. 2014). For all of these methods, an important consideration is the type of data that is required for input. Some will take relatively raw input (e.g., cDNA read counts for genes, whether functionally annotated or not) whereas others require functional or taxonomic output from annotation pipelines described above such as MG-RAST or IMG/M. In addition, these major publicly available online tools now have built-in tools for performing statistical analysis of data (Markowitz et al. 2009; Meyer et al. 2008). Regardless, normalization is a critical consideration in the preprocessing of data (see section 7.5).

12.3   Visualization

Visualization can be a powerful way of exploring metagenomic data. Though no one single software package provides the wide range of visualization tools needed for all metagenomics applications, there are a number of tools available for specific purposes. MG-RAST and IMG also now provide options for visualization of data. Several general applications for visualizing next-generation sequencing data are also indispensable for microbial community omics applications. "Strainer" provided an early example of how visualization could enable studies of genetic variation and recombination within microbial populations (Eppley et al. 2007). Integrated Genome Viewer (Robinson et al. 2011) and tablet (Milne et al. 2013), though not designed for community genomic data, can effectively visualize coverage and paired-end information, which are useful for evaluating metagenomic assemblies (see section 5.5), together with gene annotations.

Several versatile platforms specifically designed for visualizing various forms and dimensions of community omics data are now available. Anvi'o is distinguished by its versatility, interactive interface, and customizability (Eren et al. 2015). For example, you can load binning information from external applications and visualize it in parallel with coverage and taxonomic information. The anvi'server (https://anvi-server.org/) shows great promise for channeling these unique capabilities in complex visualizing into new methods of sharing data in such a way that it can be explored further, with the various forms of highly interconnected data accessible. Elviz (Environmental Laboratory Visualization) is an interactive web-based tool for visualizing metagenome data and metadata (JGI 2017), but at this point it has more limited capabilities than anvi'o. It is designed for metagenomic assemblies and allows users to plot GC content, relative abundance, phylogenetic affiliation, and length of contigs, using either publicly available or user-uploaded datasets.

A number of R-packages for visualization of meta-omics data are also now available. ShotgunFunctionalizeR offers visualization capabilities along with statistical methods for assessing functional differences between datasets (Kristiansson et al. 2009).

Krona provides a web-based visualization of taxonomic data, using a hierarchical scheme for displaying and interactively exploring metagenomic classifications (Ondov et al. 2011). Large phylogenetic trees can be visualized and edited with tools such as Dendroscope (Huson et al. 2007) and iTOL (Interactive Tree Of Life) (Letunic & Bork 2011). Meta-SMART enables exploration of protein domains of metagenomic datasets in the phylogenetic context of iTOL (Letunic et al. 2012). Visualization of metagenomic data has also played a key role in some binning applications, including anvi'o (Eren et al. 2015), emergent self-organizing maps (Dick et al. 2009; Ultsch & Moerchen 2005), and methods for nonlinear dimension reduction (Laczny et al. 2015).

12.4   Cyberinfrastructure for Environmental Omics

DNA sequencing is one of the few technologies that is advancing at a greater pace than computer processor speeds, the so-called Moore's law (Kahn 2011) (see Fig. 1.3). Yet it is beyond this issue of raw quantity of data where the true challenge lies – the multidimensional nature of environmental omics data (Fig. 12.1). Ideally, raw reads are assembled into contigs and scaffolds, which are then separated into genomic bins that represent organisms (or populations) with rich biological information about taxonomy and phenotype. Contigs are then annotated with the coordinates of genes, and the genes themselves are annotated with associated information about their biological function. In many cases, there are numerous orthologous genes, derived from related microbial strains or species, that share the same functions and that we may want to lump together or split at some level of divergence, depending on the task at hand. Each of these bins and genes is then associated with an abundance of mapped reads, transcripts, and/or peptides across samples in space and time. Each sample is linked to geochemical and environmental data that researchers want to explore in the context of all the various levels of omics data.

Tracking all of these different dimensions of data and metadata, and linking them together, remains an unresolved challenge. A project within the National Science Foundation's EarthCube program strives to synthesize cyberinfrastructural needs, challenges, and resources for microbial community omics (Gilbert et al. 2014).

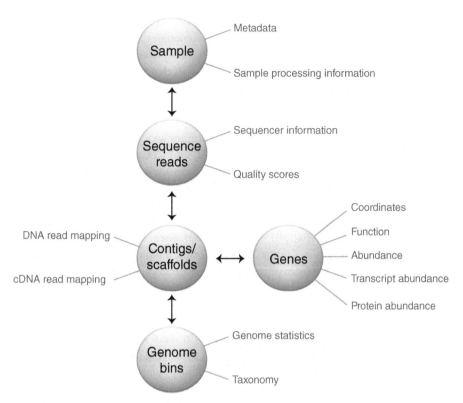

Figure 12.1 The multidimensional nature of microbial community omics data. Double arrows reflect the need to track linkages between forms of data across all dimensions. For example, such linkages are critical to study relationships between gene abundance/expression and geochemical conditions or process rates (stored as sample metadata).

12.4.1   Software Platforms for Integrated Analyses and Data Storage

As the field of environmental omics has matured, several large-scale platforms for analysis and storage of data have emerged, and a few have endured. MG-RAST (Meyer et al. 2008) and IMG (Markowitz et al. 2009) provide critical resources in terms of both analysis and databases, but have limitations in terms of breadth of analyses offered, being largely focused on annotation and downstream comparative processes. Hence, input for these pipelines is typically at the stage of contigs/scaffolds, after metagenomic assembly and binning. Due to their size, large number of users, and limited financial support, MG-RAST and IMG are also relatively inflexible to accommodate special user needs, though they increasingly provide tools that are widely useful (e.g., http://jgi.doe.gov/data-and-tools/bbtools/). CAMERA (Seshadri et al. 2007) was another "top-down" resource that served some users from the marine microbiology community well, but it was not nimble or responsive enough to many user needs.

These challenges will only multiply as the field move towards more comparative and experimental studies that involve large numbers of samples and utilize multidimensional forms of data (e.g., genes, bins, transcripts, strains across space and time), omics approaches, and parallel geochemical and environmental data.

On the upstream end of omics analyses, commercial software packages Geneious (Kearse et al. 2012) and CLC Genomics Workbench (CLC 2017) provide graphical-user interfaces for assembly, mapping reads to scaffolds and visualizing them, and conducting statistical analyses for RNAseq.

Much progress in the development of software applications has also been made on a smaller scale, by individual investigators. The anvi'o platform (Eren et al. 2015) represents a major advance in terms of flexibility and visualization. Unlike other platforms, it offers the ability to import and integrate output from other tools. For example, binning output can be easily imported and analyzed in the context of other anvi'o features. The importance of holistic analysis of community omics data – integrating and comparing single-cell genomes, metagenomes, and metatranscriptomes – was demonstrated by Eren et al. (2015) in the reanalysis of Deepwater Horizon data. This work showed that only by linking these separate sources of data could key genomic structures (most likely plasmids or phage) be discovered. These results also highlighted the value of *de novo* assembly and nucleotide composition-based binning and the limitation of taxonomy-based binning. A series of helpful tutorials and blogs are also available to support users of anvi'o (Eren 2017a). The Australian Centre for Ecogenomics has also developed a series of widely used software applications for analysis of microbial community omics data (http://ecogenomic.org/software).

Another exciting trend in the field of omics is to enable dissemination and support of user-developed applications through collaborative and open-source platforms. Here, the plant biology community provides a model with Cyverse (www.cyverse.org/), which was used as a foundation for iVirus (Bolduc et al. 2016). The US Department of Energy has also developed an open platform called kbase (https://kbase.us). A variety of smaller programming toolkits are also now available, including the Bio-Community Perl toolkit (Angly et al. 2014) and the mmgenome R package (Karst et al. 2016).

Finally, demands for hardware for metagenomics research are increasingly being met by public computing environments rather than local servers. Until it was recently retired, DIAG was a public computing environment designed specifically for genomics, with associated software and bioinformatic pipelines. JetStream provides cloud services for science and engineering in the US (https://jetstream-cloud.org/). General computing resources are also widely used; many universities now have subsidized high-performance computing facilities, and cloud computing such as Amazon also provides cost-efficient computational power.

12.5   Data and Sample Archival

Archiving data derived from molecular studies of microbial communities is crucial to ensure reproducibility and to make datasets available for reanalysis and metaanalyses. Indeed, funding agencies and journals typically require submission of data to publicly available databases. Although databases such as NCBI and IMG serve as repositories for microbial community data, they were not developed with this functionality in mind, or for community input. Thus, they do not adequately satisfy the needs of the community, especially in terms of storing and linking the many different forms of data and metadata that are inherent to microbial communities in the environment (Gilbert et al. 2014). Long-term storage and availability of samples from studies of microbial communities are also of great value but face even greater practical and cultural challenges (Cary & Fierer 2014).

As discussed above, a key challenge inherent to environmental omics data is its multidimensional nature (see Fig. 12.1). Proper archival of such data faces challenges due to its heterogeneity and complexity, but also due to the lack of incentives and appropriate resources (Brown & Tiedje 2011; Reichman et al. 2011). Hence, it is critical to develop standards and best practices for omics data and metadata, a task that has been taken on by the community-driven effort of the Genomic Standards Consortium (Field et al. 2011), which has developed the Minimal Information about a MARKer gene Sequence (Yilmaz et al. 2011). Specific communities often have specific needs, which is reflected by recommendations of the Terragenome Project (www.terragenome.org) and the ocean sciences community (Gilbert et al. 2014). Also critical for comparing data are quality control and standardization of protocols and procedures to the extent possible; this is one of the goals of the Earth Microbiome Project (Knight et al. 2012) and the microbiome quality control project (Sinha et al. 2015).

Data deluges such as the one faced by environmental genomics have been efficiently dealt with by other disciplines. For example, astronomy faced similar issues and found a solution in federated data repositories and standards for processing, interoperability, and metadata, which spurred the development of methods for analyzing and archiving data (Golden et al. 2013). Funding agencies should take a leadership role in demanding and supporting such efforts (NIH 2014).

12.6   Modeling

As discussed in section 3.1.3, the massive wave of new omics data provides new opportunities for developing and testing biogeochemical models. On the other hand, modeling is an emerging means of synthesizing omics data and generating and testing scientific hypotheses. The practice of developing models and comparing their output to real observations has high potential for synthesizing information across different spatial and temporal scales, and for bridging the microbial and biogeochemical or ecosystems sciences (Dick 2017; Mock et al. 2016). The Gordon and Betty Moore Foundation Marine Microbiology Initiative has been a strong proponent of this approach (Fuhrman et al. 2013).

To date, relatively few biogeochemical models have integrated omics data, but several approaches have been pioneered. Reed et al. (2014) used genes encoding lithotrophic energy metabolism to track the abundance of various functional groups in oxygen minimum zones. Importantly, their results demonstrated that explicit incorporation of microbes into the biogeochemical model helped to explain observed geochemical profiles. This method has also been used in conjunction with a fluid dynamics model to track microbial transport and metabolisms in the dynamic environment of deep-sea hydrothermal plumes, where the model conclusively showed that plume microbial communities must be derived from water column rather than seafloor sources (Reed et al. 2015). Louca et al. (2016) used a similar approach, but extended it to include mRNA and protein information. While many uncertainties remain regarding controls on production and degradation of mRNA and proteins in natural microbial communities, this paper highlighted the power of modeling to probe and evaluate our current understanding of these processes, and predicted that spatiotemporal datasets could provide further insights.

A key aspect of the omics-enabled, thermodynamic-kinetic, biogeochemical modeling approach is that free energy available for microbial growth from lithotrophic metabolisms can be readily predicted based on observed geochemical conditions. Further, many specific redox reactions of lithotrophic metabolisms are catalyzed by enzymes encoded by well-known and well-conserved marker genes. Thus, community composition predicted from thermodynamics can be directly compared to observations. While this appears to work well for lithotrophic metabolisms in some systems, it is not readily applicable to photosynthetic or heterotrophic communities that largely drive Earth's biogeochemical cycles.

Opportunities abound for linking genetic information to ecophysiological traits so that omics data can be integrated into ecosystem models. If marker genes for key traits in terms of nutrient and light requirements, organic substrates, cell size, and response for perturbation can be defined, then one can imagine a fruitful synthesis of gene-centric modeling with a variety of models from theoretical ecology, which have advanced substantially in the past decade (Follows & Dutkiewicz 2011; Litchman & Klausmeier 2008; Ward et al. 2014; Zomorrodi & Segre 2016). Indeed, trait-based models that are enabled and evaluated by metagenomic data promise to improve predictions about how soil microbial communities will respond to climactic perturbations such as drought (Bouskill et al. 2012; Martiny et al. 2017). Multitrophic ecosystem models (Weitz et al. 2015) are also ripe for integration with gene-centric approaches.

At a much more reductionist level, an entire field has developed around the aim of modeling cellular metabolism based on microbial genomes (Kim et al. 2012; Steuer et al. 2012). Recently, this approach was extended to whole microbial communities by assembling the species-level models from constituent community members (Zelezniak et al. 2015). This study revealed both competition and metabolite exchange between microbial groups across a wide variety of habitats. As such ecosystem and community level models begin to incorporate more details about the physiology and metabolism of specific populations (Bradley et al. 2015; Follows et al. 2007; Todd-Brown et al. 2012; Treseder et al. 2012; Wieder et al. 2015), the gap with genome-scale models, which have matured over the past 15 years (Karlsson et al. 2011; Kim et al. 2012; Zomorrodi & Maranas 2012; Zomorrodi et al. 2014), should begin to close so that each approach could benefit from the other in moving towards genome-resolved biogeochemical models of whole ecosystems. Exciting developments in this area are under way, and draw strength from direct integration of experimental data (Louca & Doebeli 2015). Transcriptomic, metabolomic, and proteomic data obtained directly from the environment, including information on enzyme concentrations, could be valuable to inform, verify, and calibrate models developed for natural communities.

12.7 Emerging Trends and Future Outlook

In 2011, Brown and Tiedje articulated a prescient vision of how metagenomics will develop in the future (Brown and Tiedje 2011). Before we can achieve the ultimate goal of a "systems biology of the biosphere" with predictive capabilities, effort and resources need to be shifted from sequencing to analysis and computation, and eventually towards experimentation and modeling. In fact, this vision is applicable not only to metagenomics but to all omics approaches, and remains largely relevant and accurate 7 years later.

Several new initiatives have the potential to tackle the computational challenge surrounding environmental omics approaches. The National Science Foundation EarthCube program began in 2011 with the goal to "enable geoscientists to address the challenges of understanding and predicting a complex and evolving Earth system by fostering a community-governed effort to develop a common cyberinfrastructure to collect, access, analyze, share and visualize all forms of data and resources, using advanced technological and computational capabilities" (Gil et al. 2014). Within EarthCube, a research coordination network is focused on environmental genomics for ocean sciences and geobiology (ECOGEO 2017).

As discussed in section 3.3 and elsewhere (Dick & Lam 2015; Druschel et al. 2014), a fundamental limitation of omics approaches is the lack of knowledge regarding the physiological and biogeochemical functions of many genes (and vice versa; which genes encode which biogeochemical processes). The magnitude of this challenge is stunning. Consider the most intensively studied bacteria on Earth, such as *E. coli*. This organism, which has been subjected to intensive lab research for decades, still harbors a substantial fraction of genes of unknown function. Now consider the vast diversity of uncultured microbes, of which we are just beginning to catch glimpses (Baker & Dick 2013; Brown et al. 2015). The gap between omics data and functional knowledge continues to widen (Galperin & Koonin 2010), and new high-throughput approaches are urgently needed in both culture-dependent (Deutschbauer et al. 2014) and culture-independent (Taupp et al. 2011) methods. Ironically, this grand challenge highlights the need for renewed efforts to develop functional metagenomic approaches, which were the focus of early metagenomic studies (Béjà et al. 2000; Riesenfeld et al. 2004). Modeling approaches and their integration with experiments and observation also hold promise as a means for navigating through vast unknowns to arrive at a systems-level understanding of microbial communities and their role in Earth, environmental, and engineered systems (Dick 2017).

Even as the lag between generation of sequence data and computational infrastructure and biochemical knowledge continues to widen with current sequencing technologies, new advances in DNA sequencing that could accelerate this disparity are on the horizon. Fortunately, certain aspects of these advances promise to ameliorate rather than exacerbate some of the primary challenges highlighted in this book. In particular, reports of ultra-long reads (hundreds of kilobases in length!) are starting to emerge. These technologies could eventually provide a path to circumvent the challenges of metagenomic assembly and binning, and thus hold potential to once again change the landscape of possibilities for environmental omics. Single-cell transcriptomics is already here for eukaryotic cells and it is likely just a matter of time before remaining complications with its application to bacterial and archaeal cells (mainly cell lysis and rRNA depletion) are resolved. Advances in structural biology, especially in computational modeling of protein structures, are already opening new perspectives on protein diversity and biology (Ovchinnikov et al. 2017) and could represent a game-changer for shedding light on metagenomic sequences, especially those of unknown function.

Finally, we may be entering an era in which it is possible to leverage knowledge of microbial community omics to manipulate microbial systems through technologies such as probiotics, phage therapy, and CRISPR. These anticipated breakthroughs make it clear that microbial community omics will remain an exciting and dynamic field for the foreseeable future. However, perhaps most exciting is that the history of this field shows that the most transformative advances are likely to come in areas and from directions that are unimaginable to us right now.

References

Angly, F. E., Fields, C. J. & Tyson, G. W. (2014) The Bio-Community Perl toolkit for microbial ecology. *Bioinformatics*, **30**, 1926–1927.

Baker, B. J. & Dick, G. J. (2013) Omic approaches in microbial ecology: charting the unknown. *Microbe*, **8**, 353–360.

Béjà, O., Aravind, L., Koonin, E. V., et al. (2000) Bacterial rhodopsin: evidence for a new type of phototrophy in the sea. *Science*, **289**, 1902–1906.

Bolduc, B., Youens-Clark, K., Roux, S., Hurwitz, B. L. & Sullivan, M. B. (2016) iVirus: facilitating new insights in viral ecology with software and community data sets imbedded in a cyberinfrastructure. *ISME Journal*, **11**, 7–14.

Bouskill, N. J., Tang, J., Riley, W. J. & Brodie, E. L. (2012) Trait-based representation of biological nitrification: model development, testing, and predicted community composition. *Frontiers in Microbiology*, **3**, 364.

Bradley, J. A., Anesio, A. M., Singarayer, J. S., Heath, M. R. & Arndt, S. (2015) SHIMMER (1.0): a novel mathematical model for microbial and biogeochemical dynamics in glacier forefield ecosystems. *Geoscientific Model Development*, **8**, 3441–3470.

Brown, C. T. & Tiedje, J. M. (2011) Metagenomics: the paths forward. In: F. de Bruijn F. (ed.), *Handbook of Molecular Microbial Ecology, vol. II: Metagenomics in Different Habitats*. John Wiley, Chichester.

Brown, C. T., Hug, L. A., Thomas, B. C., et al. (2015) Unusual biology across a group comprising more than 15% of domain Bacteria. *Nature*, **523**, 208–211.

Buttigieg, P. L. & Ramette, A. (2014a) A guide to statistical analysis in microbial ecology: a community-focused, living review of multivariate data analyses. *FEMS Microbiology Ecology*, **90**, 543–550.

Buttigieg, P. L. & Ramette, A. (2014b) *GUSTA ME*. Available at: https://sites.google.com/site/mb3gustame/ (accessed 1 November 2017).

Cary, S. C. & Fierer, N. (2014) The importance of sample archiving in microbial ecology. *Nature Reviews Microbiology*, **12**, 789–790.

CLC (2017) *CLC Main Workbench*. Available at: www.qiagenbioinformatics.com/products/clc-main-workbench (accessed 1 November 2017).

Contreras-Moreira, B. & Vinuesa, P. (2013) GET_HOMOLOGUES, a versatile software package for scalable and robust microbial pangenome analysis. *Applied and Environmental Microbiology*, **79**, 7696–7701.

Deutschbauer, A., Price, M. N., Wetmore, K. M., et al. (2014) Towards an informative mutant phenotype for every bacterial gene. *Journal of Bacteriology*, **196**, 3643–3655.

Dick, G. J. (2017) Embracing the mantra of modellers and synthesizing omics, experiments and models. *Environmental Microbiology Reports*, **9**, 18–20.

Dick, G. J. & Lam, P. (2015) Omics approaches to microbial geochemistry. *Elements*, **11**, 403–408.

Dick, G. J., Andersson, A. F., Baker, B. J., et al. (2009) Community-wide analysis of microbial genome sequence signatures. *Genome Biology*, **10**, R85.

Druschel, G. K., Dick, G. J. & Boyd, E. S. (2014) Geomicrobiology and Microbial Geochemistry 2014 Workshop Report. Available at: 10.6084/m9.figshare.3083524.v1 (accessed 1 November 2017).

ECOGEO (2017) *Oceanography and Geobiology Environmental 'Omics*. Available at: www.earthcube.org/group/oceanography-geobiology-environmental-omics (accessed 1 November 2017).

Eppley, J. M., Tyson, G. W., Getz, W. M. & Banfield, J. F. (2007) Strainer: software for analysis of population variation in community genomic datasets. *BMC Bioinformatics*, **8**, 398.

Eren, A. M. (2017a) *Anvi'o in a Nutshell*. Available at: http://merenlab.org/software/anvio/ (accessed 1 November 2017).

Eren, A. M. (2017b) *An anvi'o Workflow for Microbial Pangenomics*. Available at: http://merenlab.org/2016/11/08/pangenomics-v2/ (accessed 1 November 2017).

Eren, A. M., Esen, O. C., Quince, C., et al. (2015) Anvi'o: an advanced analysis and visualization platform for 'omics data. *PeerJ*, **3**, e1319.

Field, D., Amaral-Zettler, L., Cochrane, G., et al. (2011) The Genomic Standards Consortium. *PLoS Biol*, **9**, e1001088.

Follows, M. J. & Dutkiewicz, S. (2011) Modeling diverse communities of marine microbes. *Annual Review of Marine Science*, **3**, 427–451.

Follows, M. J., Dutkiewicz, S., Grant, S. & Chisholm, S. W. (2007) Emergent biogeography of microbial communities in a model ocean. *Science*, **315**, 1843–1846.

Fuhrman, J., Follows, M. J. & Forde, S. (2013) Meeting: Applying "-omics" data in marine microbial oceanography. *Eos*, **94**, 241.

Galperin, M. Y. & Koonin, E. V. (2010) From complete genome sequence to 'complete' understanding? *Trends in Biotechnology*, **28**, 398–406.

Gil, Y., Chan, M., Gomez, B. & Caron, B. (2014) *EarthCube: Past, Present, and Future*. Available at: www.earthcube.org/document/2014/earthcube-past-present-future (accessed 1 November 2017).

Gilbert, J. A., Dick, G. J., Jenkins, B., et al. (2014) Meeting report: Ocean 'omics science, technology and cyberinfrastructure: current challenges and future requirements (August 20–23, 2013). *Standards in Genomic Sciences*, **9**.

Golden, A., Djorgovski, S. & Greally, J. M. (2013) Astrogenomics: big data, old problems, old solutions? *Genome Biology*, **14**, 129.

Huson, D. H., Richter, D. C., Rausch, C., Dezulian, T., Franz, M. & Rupp, R. (2007) Dendroscope: An interactive viewer for large phylogenetic trees. *BMC Bioinformatics*, **8**, 460.

Ismay, C. & Kim, A. Y. (2015) *ModernDive: An Introduction to Statistical and Data Sciences in R*. Available at: https://ismayc.github.io/moderndiver-book/ (accessed 1 November 2017).

JGI (2017) *Elviz 2.0*. Available at: http://genome.jgi.doe.gov/viz/ (accessed 1 November 2017).

Kahn, S. D. (2011) On the future of genomic data. *Science*, **331**, 728–729.

Karlsson, F. H., Nookaew, I., Petranovic, D. & Nielsen, J. (2011) Prospects for systems biology and modeling of the gut microbiome. *Trends in Biotechnology*, **29**, 251–258.

Karst, S., Kirkegaard, R. & Albertsen, M. (2016) mmgenome: a toolbox for reproducible genome extraction from metagenomes. bioRxiv. Available at: www.biorxiv.org/content/early/2016/06/15/059121 (accessed 1 November 2017).

Kashtan, N., Roggensack, S. E., Rodrigue, S., et al. (2014) Single-cell genomics reveals hundreds of coexisting subpopulations in wild Prochlorococcus. *Science*, **344**, 416–420.

Kearse, M., Moir, R., Wilson, A., et al. (2012) Geneious Basic: an integrated and extendable desktop software platform for the organization and analysis of sequence data. *Bioinformatics*, **28**, 1647–1649.

Kim, T. Y., Sohn, S. B., Bin Kim, Y., Kim, W. J. & Lee, S. Y. (2012) Recent advances in reconstruction and applications of genome-scale metabolic models. *Current Opinion in Biotechnology*, **23**, 617–623.

Knight, R., Jansson, J., Field, D., et al. (2012) Unlocking the potential of metagenomics through replicated experimental design. *Nature Biotechnology*, **30**, 513–520.

Kristensen, D. M., Wolf, Y. I., Mushegian, A. R. & Koonin, E. V. (2011) Computational methods for Gene Orthology inference. *Briefings in Bioinformatics*, **12**, 379–391.

Kristiansson, E., Hugenholtz, P. & Dalevi, D. (2009) ShotgunFunctionalizeR: an R-package for functional comparison of metagenomes. *Bioinformatics*, **25**, 2737–2738.

Laczny, C. C., Sternal, T., Plugaru, V., et al. (2015) VizBin – an application for reference-independent visualization and human-augmented binning of metagenomic data. *Microbiome*, **3**, 1.

Lan, Y., Morrison, J. C., Hershberg, R. & Rosen, G. L. (2014) POGO-DB – a database of pairwise comparisons of genomes and conserved orthologous genes. *Nucleic Acids Research*, **42**, D625–632.

Letunic, I. & Bork, P. (2011) Interactive Tree of Life v2: online annotation and display of phylogenetic trees made easy. *Nucleic Acids Research*, **39**, W475–478.

Letunic, I., Doerks, T. & Bork, P. (2012) SMART 7: recent updates to the protein domain annotation resource. *Nucleic Acids Research*, **40**, D302–305.

Li, L., Stoeckert, C. J. Jr. & Roos, D. S. (2003) OrthoMCL: identification of ortholog groups for eukaryotic genomes. *Genome Research*, **13**, 2178–2189.

Litchman, E. & Klausmeier, C. A. (2008) Trait-based community ecology of phytoplankton. *Annual Review of Ecology, Evolution and Systematics*, **39**, 615–639.

Louca, S. & Doebeli, M. (2015) Calibration and analysis of genome-based models for microbial ecology. *Elife*, **4**, e08208.

Louca, S., Hawley, A. K., Katsev, S., et al. (2016) Integrating biogeochemistry with multiomic sequence information in a model oxygen minimum zone. *Proceedings of the National Academy of Sciences of the United States of America*, **113**, E5925–E5933.

Love, M. I., Huber, W. & Anders, S. (2014) Moderated estimation of fold change and dispersion for RNA-seq data with DESeq2. *Genome Biology*, **15**, 550.

Markowitz, V. M., Mavromatis, K., Ivanova, N. N., Chen, I. M., Chu, K. & Kyrpides, N. C. (2009) IMG ER: a system for microbial genome annotation expert review and curation. *Bioinformatics*, **25**, 2271–2278.

Martiny, J. B. H., Martiny, A. C., Weihe, C., et al. (2017) Microbial legacies alter decomposition in response to simulated global change. *ISME Journal*, **11**, 490–499.

McMurdie, P. J. & Holmes, S. (2013) phyloseq: an R package for reproducible interactive analysis and graphics of microbiome census data. *PLoS One*, **8**, e61217.

Meyer, F., Paarmann, D., d'Souza, M., et al. (2008) The metagenomics RAST server – a public resource for the automatic phylogenetic and functional analysis of metagenomes. *BMC Bioinformatics*, **9**, 386.

Milne, I., Stephen, G., Bayer, M., et al. (2013) Using Tablet for visual exploration of second-generation sequencing data. *Briefings in Bioinformatics*, **14**, 193–202.

Mock, T., Daines, S. J., Geider, R., et al. (2016) Bridging the gap between omics and earth system science to better understand how environmental change impacts marine microbes. *Global Change Biology*, **22**, 61–75.

NIH (2014) *NIH Genomic Data Sharing Policy*. Available at: http://grants.nih.gov/grants/guide/notice-files/NOT-OD-14-124.html (accessed 1 November 2017).

Ondov, B. D., Bergman, N. H. & Phillippy, A. M. (2011) Interactive metagenomic visualization in a Web browser. *BMC Bioinformatics*, **12**, 385.

Ovchinnikov, S., Park, H., Varghese, N., et al. (2017) Protein structure determination using metagenome sequence data. *Science*, **355**, 294–298.

Parks, D. H., Tyson, G. W., Hugenholtz, P. & Beiko, R. G. (2014) STAMP: statistical analysis of taxonomic and functional profiles. *Bioinformatics*, **30**, 3123–3124.

Primer-E. (2017) *Multivariate Statistics for Ecologists*. Available at: www.primer-e.com/ (accessed 1 November 2017).

Reed, D. C., Algar, C. K., Huber, J. A. & Dick, G. J. (2014) Gene-centric approach to integrating environmental genomics and biogeochemical models. *Proceedings of the National Academy of Sciences of the United States of America*, **111**, 1879–1884.

Reed, D. C., Breier, J. A., Jiang, H., et al. (2015) Predicting the response of the deep-ocean microbiome to geochemical perturbations by hydrothermal vents. *ISME Journal*, **9**, 1857–1869.

Reichman, O. J., Jones, M. B. & Schildhauer, M. P. (2011) Challenges and opportunities of open data in ecology. *Science*, **331**, 703–705.

Riesenfeld, C. S., Goodman, R. M. & Handelsman, J. (2004) Uncultured soil bacteria are a reservoir of new antibiotic resistance genes. *Environmental Microbiology*, **6**, 981–989.

Robinson, J. T., Thorvaldsdottir, H., Winckler, W., et al. (2011) Integrative genomics viewer. *Nature Biotechnology*, **29**, 24–26.

Seshadri, R., Kravitz, S. A., Smarr, L., Gilna, P. & Frazier, M. (2007) CAMERA: a community resource for metagenomics. *PLoS Biology*, **5**, e75.

Simpson, G., Oksanen, J., Solymos, P., Stevens, H., Legendre, M. & Wagner, F. G. (2017) *Vegan*. Available at: http://vegan.r-forge.r-project.org/ (accessed 1 November 2017).

Sinha, R., Abnet, C. C., White, O., Knight, R. & Huttenhower, C. (2015) The microbiome quality control project: baseline study design and future directions. *Genome Biology*, **16**, 276.

STAMPS (2017) *STAMPS Schedule*. Available at: https://stamps.mbl.edu/index.php/Schedule (accessed 1 November 2017).

Steuer, R., Knoop, H. & Machne, R. (2012) Modelling cyanobacteria: from metabolism to integrative models of phototrophic growth. *Journal of Experimental Botany*, **63**, 2259–2274.

Taupp, M., Mewis, K. & Hallam, S. J. (2011) The art and design of functional metagenomic screens. *Current Opinion in Biotechnology*, **22**, 465–472.

Todd-Brown, K. E. O., Hopkins, F. M., Kivlin, S. N., Talbot, J. M. & Allison, S. D. (2012) A framework for representing microbial decomposition in coupled climate models. *Biogeochemistry*, **109**, 19–33.

Treseder, K. K., Balser, T. C., Bradford, M. A., et al. (2012) Integrating microbial ecology into ecosystem models: challenges and priorities. *Biogeochemistry*, **109**, 7–18.

Ultsch, A. & Moerchen, F. (2005) *ESOM-Maps: Tools for Clustering, Visualization, and Classification with Emergent SOM*. Technical Report. Department of Mathematics and Computer Science, University of Marburg, Germany, p. 46.

Ward, B. A., Dutkiewicz, S. & Follows, M. J. (2014) Modelling spatial and temporal patterns in size-structured marine plankton communities: top-down and bottom-up controls. *Journal of Plankton Research*, **36**, 31–47.

Weitz, J. S., Stock, C. A., Wilhelm, S. W., et al. (2015) A multitrophic model to quantify the effects of marine viruses on microbial food webs and ecosystem processes. *ISME Journal*, **9**, 1352–1364.

Wieder, W. R., Allison, S. D., Davidson, E. A., et al. (2015) Explicitly representing soil microbial processes in Earth system models. *Global Biogeochemical Cycles*, **29**, 1782–1800.

Wurtzel, O., Dori-Bachash, M., Pietrokovski, S., Jurkevitch, E. & Sorek, R. (2010) Mutation detection with next-generation resequencing through a mediator genome. *PLoS One*, **5**, e15628.

Yilmaz, P., Gilbert, J. A., Knight, R., et al. (2011) The genomic standards consortium: bringing standards to life for microbial ecology. *ISME Journal*, **5**, 1565–1567.

Zelezniak, A., Andrejev, S., Ponomarova, O., Mende, D. R., Bork, P. & Patil, K. R. (2015) Metabolic dependencies drive species co-occurrence in diverse microbial communities. *Proceedings of the National Academy of Sciences of the United States of America*, **112**, 6449–6454.

Zomorrodi, A. R. & Maranas, C. D. (2012) OptCom: a multi-level optimization framework for the metabolic modeling and analysis of microbial communities. *PLoS Computational Biology*, **8**, e1002363.

Zomorrodi, A. R. & Segre, D. (2016) Synthetic ecology of microbes: mathematical models and applications. *Journal of Molecular Biology*, **428**, 837–861.

Zomorrodi, A. R., Islam, M. M. & Maranas, C. D. (2014) d-OptCom: dynamic multi-level and multi-objective metabolic modeling of microbial communities. *ACS Synthetic Biology*, **3**, 247–257.

Index

Italics indicate page numbers referring to a figure.
Bold indicates page numbers referring to a table or box.

Genomic Approaches in Earth and Environmental Sciences, First Edition. Gregory Dick.
© 2019 John Wiley & Sons Ltd. Published 2019 by John Wiley & Sons Ltd.